A Cabinet of Mathematical
Curiosities at Teachers College:

DAVID EUGENE SMITH'S
COLLECTION

DIANE R. MURRAY

Docent Press
Boston, Massachusetts, USA
www.docentpress.com

Docent Press publishes books in the history of mathematics and computing about interesting people and intriguing ideas. The histories are told at many levels of detail and depth that can be explored at leisure by the general reader.

Cover design by Brenda Riddell, Graphic Details. Typeset with TeX.

© Diane R. Murray 2013

All rights reserved. No part of this book may be reproduced or utilized in any form or by any means, electronic or mechanical, including photocopying and recording, or by any information storage and retrieval system, without permission in writing from the author.

For my mother

Contents

	Acknowledgements	x
1	Introduction	1
	1.1 David Eugene Smith	1
	1.2 Organization of the Book	2
	1.3 Previous Work	4
	1.4 The History of a Collection	5
	1.5 References	6
2	Smith Travels the World	7
	2.1 Education and Early Career	9
	2.2 Italy — 1904	14
	2.3 Burma, India, Sri Lanka, Japan, and China — 1907	18
	2.4 Sumatra, Thailand — 1930	25
	2.5 Persia — 1933	28
	2.6 References	31
3	Early History of Teachers College and the Educational Museum	33
	3.1 Industrial Education Association	33
	3.2 Teachers College	37
	3.3 The Educational Museum	40
	3.4 References	51
4	Step Inside Smith's Collection in the Educational Museum	53
	4.1 D. E. Smith's Collection in His Own Words	53
	4.2 Printed Material	56
	4.3 Manuscripts	58
	4.4 Portraits and Medallions	61
	4.5 Autographs	62

	4.6	Instruments	65
	4.7	Two Exhibition Pamphlets	75
	4.8	References	76
5	Smith's Collection on the Move	79
	5.1	References	85
6	A Collection Without Walls	89
	6.1	The Lantern Slides	90
	6.2	Texts	92
	6.3	Teaching and Mentoring	95
	6.4	References	97
7	Columbia University Receives a Gift	101
	7.1	Smith's Donation	101
	7.2	Smith's Collection Today	107
	7.3	References	109
8	Conclusion	111
	8.1	References	116
A	"Illustrations for Lectures on the History of Mathematics" (1907)	117
B	Lantern Slides	141
C	Two Pamphlets	299
D	List of Portraits in Smith's Collection	313
E	Catalog of Smith's Collection of Mathematical Instruments	333
F	Countries Visited by David Eugene Smith	371
	Index	373

List of Figures

2.0.1	Smith as a young child.	8
2.1.1	David Eugene Smith at age 21 upon his graduation from Syracuse University.	9
2.1.2	David Eugene Smith at age 35 teaching at Michigan State Normal College in Ypsilanti.	11
2.1.3	Smith (center), Helen McAleer (front left), and Clara Jewett (far right) in 1937.	12
2.1.4	a. Flyer sent to loyal customers of Helen Jewett McAleer, Smith's niece, in 1930. a. Smith in McAleer's shop in 1934. It was common for Smith to visit the family home to get away from New York City. He also gave presentations on the items in McAleer's gallery from time to time.	13
2.2.1	Third letter in a collection of 124 letters from Prince Boncompagni to Ferdinando Jacoli.	16
2.2.2	Libri's *History of Mathematics in Italy* (1835). The notes, among a few other pieces, are placed in the front of the text.	17
2.5.1	Photograph sent to Miss Bertha M. Frick during his travels through Persia in 1933.	28
3.0.1	Grace Hoadley Dodge (1856–1914). Image provided courtesy of the Gottesman Libraries at Teachers College, Columbia University.	34
3.1.1	Nicholas Butler (1862–1947). Image provided courtesy of the Gottesman Libraries at Teachers College, Columbia University.	36
3.2.1	Arithmetic Class. Seventh Grade. No. 9 University Place, Teachers College, 1893.	39
3.2.2	James Earl Russell (1898–1926). Image provided courtesy of the Gottesman Libraries at Teachers College, Columbia University.	40

List of Figures

3.3.1	Educational Museum of Teachers College. Image is provided courtesy of the Gottesman Libraries at Teachers College, Columbia University.	41
3.3.2	Photograph of the Department of Mathematics exhibit room, circa 1903.	48
3.3.3	Photographs of the Department of Mathematics exhibit room. a. facing towards the door. b. facing opposite direction.	50
4.1.1	Photograph of the Department of Mathematics exhibit room, circa 1903.	55
4.2.1	Paolo Casati's *Fabrica et uso del compasso di proportione* (1685). a. Title page. b. Sample page including decimal fractions.	58
4.2.2	Tonstall's *De Arte Supputandi* (1529) with the autograph of Thomas Digges. On the left hand page, George A. Plimpton, who presented this text to Smith, signed it.	58
4.2.3	Gemma Fresius' *Arithmeticæ practicæ methodus facilis* (1540) title page.	59
4.2.4	Smith's bookplate (1907), modeled after Fresius' bookplate where Smith used his own image instead.	59
4.3.1	Unpublished manuscript of Galileo's life. MS 520.1800. Description: "Manuscript written in the late 18th century. Apparently the author was a contemporary of Galileo. Numerous errors corrected in red, perhaps by Ferdinando Jacoli, to whom this ms. probably belonged."	61
4.4.1	Portrait of Guillaume de l'Hôpital.	62
4.5.1	Letter from Charles Lutwidge Dodgson (Lewis Carroll) to the Mathematical Editor of *The Educational Times*.	64
4.6.1	a. Armillary sphere. Italy, circa 1550. b. Quadrant. Italian, ivory with brass cover, early 19th century. c. Astrolabe. Italian, signed by Bernard Sabeus, 1558.	66
4.6.2	Ancient Roman compass, about the beginning of the Christian era.	66
4.6.3	Examples of dice. a. Roman. Gypsum. b. Roman. Bone. c. A hexagonal prism. Bone. d. Three parts of the die.	68
4.6.4	Examples of tesseræ with descriptive labels.	69
4.6.5	"Japanese sangi sticks used in solving equations in the Old Japanese algebraic system. Purchased in Kyoto, Japan, 1907."	69
4.6.6	English tally sticks of 1296.	71

List of Figures

4.6.7	"Nuremberg, signed by Hans Tröchel, 1603. Ivory, with string gnomon horizontal dial and pin gnomon for vertical dial."	72
4.6.8	"Cubical sundial. Bavarian, 18th century. Horizontal and vertical. North, south, east, and west."	72
4.6.9	Ramsden Telescope.	73
4.6.10	Representations of the Magic Square.	74
7.2.1	Part of Case X at the 2002 Exhibit. Includes nests of weights and money changer's balances.	108
8.0.1	Department of Mathematics Office, circa 1910, with prints of celebrated mathematicians on the back wall.	113

Acknowledgements

First and foremost, I would like to thank my mentor Dr. Bruce R. Vogeli, who has always believed in me and who I consider a motivator, beloved friend, and "grandfather." It is through Dr. Vogeli that I started this research when he handed me boxes filled with lantern slides from David Eugene Smith's collection. Smith was quite a traveller and storyteller—an "Indiana Jones" for mathematics, if you will. I was privileged to hear Dr. Vogeli's numerous adventures throughout his life and consider him as an "Indiana Vogeli." He is a magnificent, gifted, and generous professor, who is still actively teaching and travelling, now going into his mid-eighties! Along with Dr.Vogeli, my other advisors at Teachers College were Drs. Alexander P. Karp and Erica N. Walker. These two brilliant professors provided invaluable insight and knowledge with historical studies and powerful writing. The wonderful personalities and sense of humors of Drs. Vogeli, Karp, and Walker have filled my heart with the best memories and I will cherish them always. These three superb professors have always made me feel like a colleague and I am blessed to still have them in my life.

Three other important people who aided me in this study are Jennifer B. Lee, the curator of the David Eugene Smith collection at the Rare Book and Manuscript Library (RBML) of Columbia University. She was always available to guide me to the item I needed at the time and chat with me about Smith and his collection. It is through her and the RBML that the photographs of Smith's archives and his collection are published here. Allen Foresta, senior librarian at the Gottesman Libraries of Teachers College, was always enthusiastic in my searches and I consider him a fellow protector of historical Teachers College matters. Eileen Donoghue, professor at College of Staten Island, and I share a love of David Eugene Smith. Through a few conversations and letters, along with sharing some lantern slides, Eileen shed some light on my topic and I am fortunate to say that I know such a dignified woman.

I feel lucky to still have such a close connection with my undergraduate professors at Manhattanville College. Drs. Edward Schwartz, Phyllis Lefton, Gerard Kiernan, Norman Bashias, Mirela Djordjevic, Arnold Koltun, and Michelle Longhitano provided the strong foundation in mathematics and computer science as well as inspiration for me to continue my education to the doctoral level. I must distinguish Drs. Schwartz, Lefton, and Kiernan who have continuously been interested in my current work and encourage me to always do my best.

Acknowledgements

This book would not be as it is now without the wonderful guidance of Scott Guthery, design of Brenda Riddell, and encouragement of Mary Cronin of Docent Press.

My family has always been my support system throughout my life. Without my mother, Linda, my father, Gordon, my sister, Trisha, and my fiancé, Adam, I would have not been able to accomplish this goal. I am sincerely grateful and blessed to have them in my life, especially my mother who allowed me the freedom to focus solely on my research. I also have to thank my two "babies"—Stewie and Spot—four-year-old rat terrier brothers, who curled up with me as I wrote this study.

Chapter 1

Introduction

1.1 David Eugene Smith

Adventurer, collector, and world traveler are not typically used to describe mathematics professors. David Eugene Smith (1860–1944), however, was entirely these and more.

Smith is best known as a dedicated mathematics professor, textbook author, developer of mathematics education programs, and editor. His contributions to the field include the organization and leadership of the *International Commission on the Teaching of Mathematics* (*ICMI*) and co-founding a respected journal devoted to the philosophy, history, and expository treatment of mathematics, *Scripta Mathematica*.

Smith's major publications were on the history of mathematics. Among them were the acclaimed *Rara Arithmetica* (1908), *The Hindu-Arabic Numerals* (1911), *A History of Japanese Mathematics* (Smith & Mikami, 1914), *Number Stories of Long Ago* (1919), *The History of Mathematics Volume I and II* (1923; 1925), *A History of Mathematics in America Before 1900* (Smith & Ginsburg, 1934), *The Poetry of Mathematics and Other Essay* (1934), *Numbers and Numerals* (Ginsburg & Smith, 1937), and *The Wonderful*

Wonders of One-Two-Three (1937). Bertha M. Frick provides a detailed bibliography of the writings of David Eugene Smith in the first issue of *Osiris* in 1936 (Frick, 1936a).

Throughout his career however, Smith was most passionate about building a collection that encompassed a global history of mathematics. He traveled the world to acquire rare books, artifacts, and manuscripts that would document the evolution of mathematical discovery and practice across widely different cultures. Smith was not collecting these treasures solely for his own personal satisfaction. He was a pioneer in integrating historical artifacts into the teaching of mathematics. Smith believed that mathematics education would be enhanced through the study and appreciation of historical pieces. This was a belief he espoused not only for his own students but also for the training of mathematics teachers and their pupils throughout the United States.

Smith's unique collection became a centerpiece of the Educational Museum of Teachers College in New York when Smith joined the faculty in 1901. This book tells the story of Smith's adventures in amassing, exhibiting, and then disseminating information about the historical books, manuscripts, portraits, and rare instruments related to mathematics he collected during his career. It provides a detailed profile of the organization of the Smith collection at its high point early in the twentieth century and describes the fate of these materials over the past one hundred years as Smith's legacy searched for a permanent home.

1.2 Organization of the Book

Part One of the book provides a candid profile of Smith as an adventurous collector, using accounts from Smith's unpublished letters and manuscripts housed in the Rare Book and Manuscript Library of Columbia University, New York. It offers a contemporary view of the original scope of the mathematical holdings of the Educational Museum and of the educational applications envisioned by Smith based on key early documents. Source materials include Smith's 1907 publication, "Illustrations for lectures on the history of mathematics," *Educational Museum: Teachers College, Columbia University, N.Y.*, and an examination of two very rare publications from the Educational Museum and the Department of Mathematics of Teachers College. Smith understood that only people in the New York City area or travelers with ample means could view his collection on location at the Educational Museum. His solution was to take advantage of the technology of his time to produce a set of lantern slides illustrating the collection's most important objects. Copies of these lantern slides were distributed to interested schools and educators throughout the country (Smith, 1907).

Smith traveled the world for forty years, crossing the Atlantic by ship over eighty times (Lawler, 1938). Part One opens with a recounting of Smith's adventures as he traveled through Italy in 1904, Burma, India, Sri Lanka, Japan, and China in 1907, Sumatra and Thailand in 1930, and Persia in 1933. The details of Smith's journeys are taken from his own account, in what he called a "little folding book." His notes are filled with stories of meeting princes in royal palaces, congregating in mud huts with natives, and bargaining with booksellers. Through these personal accounts, the story of Smith's collection comes to life.

Subsequent chapters in Part One describe the varied components of the collection in its original institutional setting at Teachers College, then follows Smith's collection as it moved from the Educational Museum of Teachers College to multiple temporary locations before it reached its final home at Columbia University. An important part of this story is the educational applications that Smith found as he integrated historical mathematics artifacts into his printed works, student research, and his lectures as a professor at Teachers College. Columbia University is where Smith decided to leave his collection for future generations to enjoy. The story brings the collection to the present day, over a century from when it was first established, including a discussion of how Smith's vision of integrating the history and teaching of mathematics has influenced mathematics teaching today.

Part Two and the Appendixes illustrate Smith's collection through reproductions of the remaining lantern slides and provide insight into how the objects in the collection were organized while at Teachers College. The illustrations are reproduced courtesy of the David Eugene Smith Archives at Columbia University's Rare Book and Manuscript Library. This archive contains printed materials, manuscripts, portraits and medallions, autographs, and mathematical instruments from his collection. Since Smith's entire collection is no longer on exhibit, this book provides a special look at the collection. Guided by Smith's own words we see the highlights of the collection and what he most valued.

1.3 Previous Work

Despite many scholarly studies of Smith and his role in mathematics education, the story of Smith's collecting activities and the collection itself has not previously been covered in detail. Three previous dissertations that focused specifically on Smith, *Educational Contributions of David Eugene Smith* written by Weatha Gale McNeil in 1986, *The Origins of a Professional Mathematics Education Program at Teachers College* written by Eileen Donoghue in 1987, and the author's own dissertation which this book is based (Murray, 2012a) analyze related aspects of Smith's career. McNeil's dissertation is an historical study of the role that Smith played in the evolution of mathematics education while,

Donoghue's dissertation is an historical account of Smith's involvement in the development of a model mathematics education program. These two studies mention that Smith was an avid collector but do not analyze the Educational Museum nor Smith's personal collection. It is imperative that this part of Smith's history not be lost.

An early dissertation, completed at Teachers College by Benjamin R. Andrews in 1909, entitled *Museums of Education: Their History and Their Use* discusses educational museums throughout the country and internationally (Andrews, B. R. 1909). Andrews covers the Educational Museum in detail in his study, along with brief mentions of the Department of Mathematics and Smith's personal collection. Since Andrews was a former supervisor of the Educational Museum, his research about the museum is essential to this book and will be referenced throughout the text. However, Andrews' study was completed during the middle years of the Educational Museum and thus cannot provide any perspective on the closing of the Educational Museum or how the items of the museum were eventually dispersed.

1.4 The History of a Collection

The items in Smith's collection were a major resource for mathematics teachers, students, and researchers. Components of the collection were also the inspiration for numerous research publications both domestic and international. Many of the portraits of mathematicians, historical documents, mathematical manuscripts, and instruments in Smith's collection are still used in researching and teaching the history of mathematics.

After a brief biographical introduction, our narrative begins by following some of Smith's remarkable journeys to obtain the items in his collection. We then sketch the turbulent context in which

Smith struggled to create a museum to house his collection. The core of the book is a description of the collection itself and how Smith used the collection in his own research and teaching and how he worked diligently to enable others to access the material in the collection. The book concludes with a short summary of the state of the collection today.

1.5 References

Andrews, B. R. (1909). *Museums of education: Their history and their use* (Doctoral dissertation). Teachers College, Columbia University, New York.

Donoghue, E. F. (1987). *The origins of a professional mathematics education program at Teachers College* (Doctoral dissertation). Teachers College, Columbia University, New York.

Frick, B. M. (Comp.). (1936a). *Bibliography of the critical, historical, and pedagogical writings of David Eugene Smith. Osiris* *I*(1).

Lawler, T. B. (1938). *Seventy years of textbook publishing: A history of Ginn and Company*. New York: Ginn and Company.

McNeil, W. G. (1986). *Educational contributions of David Eugene Smith* (Doctoral dissertation). Teachers College, Columbia University, New York.

Murray, D. R. (2012a). *A cabinet of mathematical curiosities at Teachers College: David Eugene Smith's collection*. Teachers College, Columbia University, New York.

Smith, D. E. (1907). Illustrations for lectures on the history of mathematics. *Educational museum: Teachers College Columbia University, N.Y.*

Chapter 2

Smith Travels the World

From a young age, David Eugene Smith traveled the world, loved museums, and envisioned himself as a future collector. Smith was born on January 21, 1860 in Cortland, New York to a middle class family (Figure 2.0.1). Smith's father was a lawyer and his mother was the daughter of a physician. He, along with his three siblings, learned Greek and Latin at home from their mother (Dauben, 2002). His parents instilled in him a respect for education, love of the arts and the classics, and a deep interest in books, collecting, and travel. As a family they visited museums and historic landmarks, encouraging his life-long appreciation of history. As a young man of nineteen, he traveled to Europe for a two-month adventure and then to Central and South America when he was twenty-three years old (Donoghue, 1998).

After he had read Augusta J. Evans' *St. Elmo* (1866), young Smith dreamed of how he would house his own historical collections in beautifully furnished rooms. The following excerpt from Evans' book inspired his imagination:

> On a *verd antique* table lay a satin cushion holding a vellum MS., bound in blue velvet, whose uncial letters were written in purple ink, powdered with gold-

Figure 2.0.1: Smith as a young child. David Eugene Smith Professional Papers, 1860–1945 (Box 121), Rare Book and Manuscript Library, Columbia University.

dust, while the margins were stiff with gilded illuminations;... A small Byzantine picture... hung over an étagère,... where lay a leaf from Nebuchadnezzar's diary, one of those Babylonian bricks... Several handsome rosewood cases were filled with rare books—two in Pali—centuries old; and moth-eaten volumes and valuable MSS.—some in parchment, some bound in boards —recalled the days of astrology and alchemy. (Simons, 1945, p. 41)

Smith was determined that someday he would have a room just like that—and in fact, he eventually did.

Figure 2.1.1: David Eugene Smith at age 21 upon his graduation from Syracuse University. David Eugene Smith Professional Papers, 1860–1945 (Box 121), Rare Book and Manuscript Library, Columbia University.

2.1 Education and Early Career

Smith's family background and early exposure to the arts and classics help to explain why his career plans did not initially focus on teaching mathematics. His higher education began at the State Normal School in Cortland. He continued his studies at Syracuse University where he concentrated on art, classical languages, and Hebrew (Dauben, 2002). Smith graduated in 1881 (Figure 2.1.1) and began studying law, first informally in his father's office and then formally at Syracuse University. He was admitted to the New York Bar in 1884.

Once he started practicing law, however, Smith soon realized that the legal profession was not his passion. He accepted an offer to be a mathematics instructor at his undergraduate alma mater, the State Normal School in Cortland, a position which became the foundation of his long teaching career. At the same time, however, he renewed his pursuit of graduate education at

Syracuse University, eventually receiving a Ph.M. degree in 1884 and a Ph.D. in art history in 1887. Donoghue notes:

> In the summer of 1885, he visited Europe again, this time in search of mathematical texts to add to Cortland's library and portraits of past mathematical giants to illustrate his lectures on mathematics and its history. (Donoghue, 1998, p. 360)

In 1891 Smith became the chair of the Department of Mathematics at the Michigan State Normal College in Ypsilanti (Figure 2.1.2). During his time at the Normal College, he "reformed the department, organized a prototype program to train mathematics teachers, and built an impressive mathematics library of over 700 volumes" (Donoghue, 1998, p. 361). In 1895, Ginn & Company published Smith's first textbook, *Plane and Solid Geometry*, coauthored with Wooster Woodruff Beman. Only one year later Ginn published Smith's second text, *History of Modern Mathematics*. Even at this early stage in his career Smith's publications and collections were acknowledged internationally, specifically in Germany, where Dr. Rudolf Knilling wrote an article for a German magazine devoted entirely to Smith (Lawler, 1938).

In 1898 Smith was selected to be the principal of the State Normal School at Brockport. This career choice brought him back to New York where Smith became acquainted with George Arthur Plimpton (1855–1936), president of the publishing firm Ginn & Company. Due to their shared interest in collecting historical manuscripts, Plimpton invited Smith to his home in New York City in 1900 to view his textbook collection and to get Smith's opinion on some of the items. This meeting marked the beginning of a friendship that lasted for the rest of both of their lives. The visit with Plimpton, the excitement of New York City, and the possibilities that Columbia University and its Teachers College could provide, induced Smith to accept the chair of the Depart-

2. Smith Travels the World

Figure 2.1.2: David Eugene Smith at age 35 teaching at Michigan State Normal College in Ypsilanti. David Eugene Smith Professional Papers, 1860–1945 (Box 121), Rare Book and Manuscript Library, Columbia University.

ment of Mathematics at Teachers College offered to him by Dean James Earl Russell in 1901(Russell, 1937).

After his marriage in 1887 to Fannie Taylor, Smith made it a point to include his family on his collecting adventures. His typical travel companions were his wife, Fannie Taylor, and later his sister, Mrs. Clara L. Jewett, along with her daughter, Mrs. Helen Jewett McAleer (Figure 2.1.3). McAleer in particular shared Smith's deep interest in traveling and education. She also was an avid collector and owned her own gallery, Helen Jewett's Little Bungalow Shop, which was located in their family home in Cortland, New York (Figure 2.1.4). During their visits to Europe, Japan, China, India, the Far East, and the Mediterranean, Smith and his traveling companions studied the school systems and met with local educational leaders. Thus Smith's trips contributed to his expertise on the teaching of mathematics abroad and provided him with opportunities to share information and lecture in the countries that he visited about mathematical education practices in America (Smith, n.d.a).

Figure 2.1.3: Smith (center), Helen McAleer (front left), and Clara Jewett (far right) in 1937. David Eugene Smith Professional Papers, 1860–1945 (Box 121), Rare Book and Manuscript Library, Columbia University.

In 1936, Smith reflected on some of his journeys to Italy, Burma, India, Sri Lanka, Japan, China, Sumatra, Thailand, and Persia (Smith, 1936a). These notes on his travels focused on his collecting experiences and interesting incidents. The source for these recollections is contained in twenty-four typed pages titled "little folding book" (Smith, 1936a). These notes were never formally published during Smith's lifetime. The excerpts in this chapter were published in the author's dissertation and publication in the *International Journal for the History of Mathematics Education* (Murray, 2012b) and are included here with permission of Columbia University Libraries.

Scattered throughout the text of the "little folding book" are remarks from Bertha Frick who was the curator of the George A. Plimpton, David E. Smith, and Samuel S. Dale Libraries at Columbia University beginning in 1937. The narratives in these

2. Smith Travels the World

a. Flyer sent to loyal customers of Helen Jewett McAleer, Smith's niece, in 1930.

b. Smith in McAleer's shop in 1934. It was common for Smith to visit the family home to get away from New York City. He also gave presentations on the items in McAleer's gallery from time to time.

Figure 2.1.4: David Eugene Smith Professional Papers, 1860–1945 (Box 32, Box 121), Rare Book and Manuscript Library, Columbia University.

notes seem to have been transcribed from interviews with Smith though the interviewer is not identified; however, a memoriam dedicated to Smith in the 1945 *Bulletin of the American Mathematical Society* written by Lao Genevra Simons includes some direct quotations from the narratives. This might indicate that Simons was the interviewer. Indeed, Simons was a close friend of Smith since a September 21, 1944 letter from Jekuthiel Ginsburg to Bertha M. Frick in response to Smith's passing, states: "with the exception of the members of his immediate family and Professor Lao G. Simons, we have more reasons to mourn his loss than anybody else" (Ginsburg, 1944, p. 1). These descriptions of Smith's pursuit of mathematical curiosities and historical treasures over a period of several decades provide insight into how his collection was formed.

2.2 Italy —1904

During a 1904 trip to Italy, Smith was occupied with acquiring books and artifacts for both himself and George A. Plimpton (Smith, 1936a). The trip that summer began with a visit to Leo S. Olschki, editor of the Olschki Publishing House in Florence, who at that time, according to Smith, "probably had the largest and best collection of medieval manuscripts and incunabula of any dealer in Europe" (Smith, 1936a, p. 14). After Florence, Smith went to Venice to peruse the bookshop of Olschki's son-in-law, one Mr. Rosen. Smith did not find anything of interest in the shop, but recounts that "after I left, and was walking along the piazza, I heard [Mr. Rosen] calling to me and I stopped and he said that it had just occurred to him that Professor Jacoli had a large library on the history of Italian mathematics and he wished to dispose of it" (Smith, 1936a, p. 14). Professor Jacoli's son had recently committed suicide by drowning himself in the Grand Canal when his fiancé abandoned him. Sadly, Professor Jacoli and his wife lived in an apartment overlooking the Grand Canal. They no longer wished to live there or in Venice because of their son's tragic death and planned to go back to their home in Modena (Smith, 1936a).

Ferdinando Jacoli was a professor at the naval college in Venice and had been a great friend of Prince Baldassarre Boncompagni (1821–1894). Boncompagni was an Italian historian of mathematics and considered to be the "most prominent and influential figure in this field" (Dauben, 2002, p. 80). Using his own printing press and funds, he created the renowned *Bulletino di bibliografia e storia delle scienze matematiche* (1868–1887). Jacoli had written a number of articles for the *Bulletino* which Smith had read. Mr. Rosen arranged for Smith to meet with Jacoli in his apartment library, which consisted of "a mass of piles of books of all sorts"

(Smith, 1936a, p. 15). Smith spent several hours looking them over, and asked Jacoli how much he would like for the entire lot. As Smith recounts, "the bargaining continued for a couple days, and eventually we came to terms" (Smith, 1936a, p. 15).

In Smith's view, among the most important items from Jacoli's collection were a portfolio of letters from Boncompagni to Jacoli (Figure 2.2.1), a complete collection of the *Bulletino*, and an exceedingly rare copy of the first impression of Guillaume Libri's *History of Mathematics in Italy* (Figure 2.2.2). Smith described the rarity of the latter as follows:

> The day on which the printing was finished for Volume I of that book in Paris, Libri stopped at the printing office and took a few copies of the book to his home. An hour later the printing establishment was burned, and every copy of the History was destroyed except the few that he had carried home. One of these, Libri gave to Jacoli, who was one of his greatest friends. Libri went to work at once to revise the book, and there are a large number of corrections in his handwriting and in Jacoli's. It was not until four years later [1838] that the so-called first edition was printed and put on the market. (Smith, 1936a, p. 15)

It is interesting to note that around 1897, Boncompagni's entire personal library had been put on the market. At that time, Smith had tried to find someone to help him purchase the collection, but he was still in Ypsilanti and had been unsuccessful. During this 1904 trip he did manage to purchase a good portion of Boncompagni's collection from various dealers in Florence and Rome (Smith, 1936a).

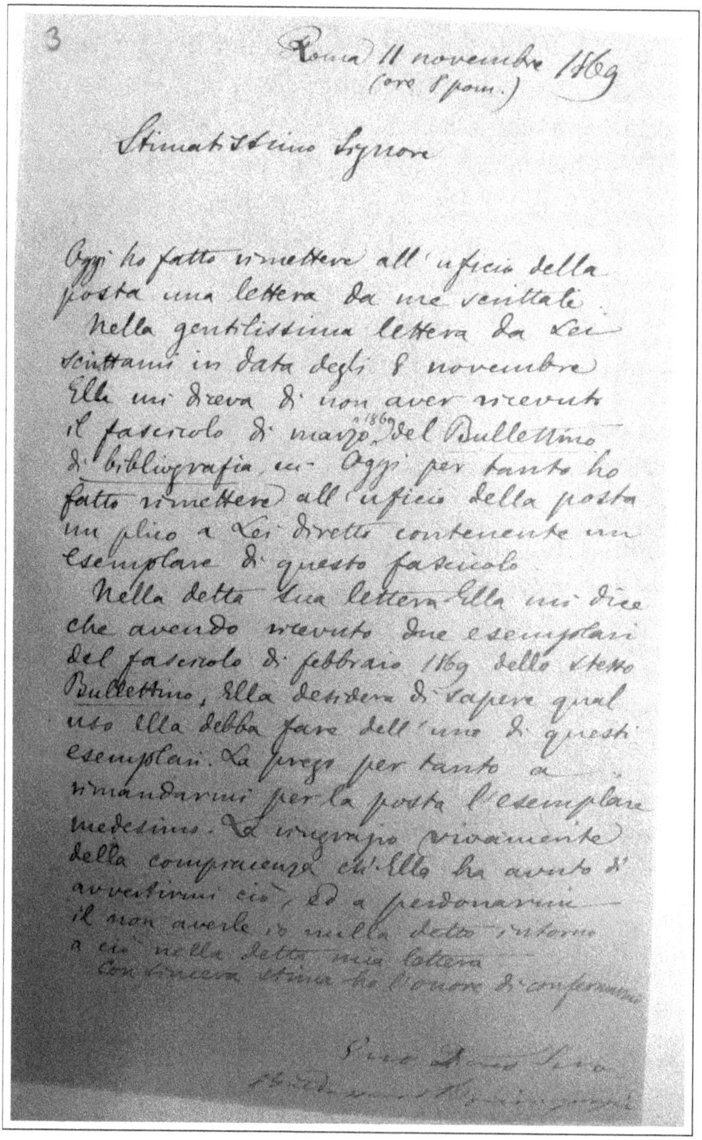

Figure 2.2.1: Third letter in a collection of 124 letters from Prince Boncompagni to Ferdinando Jacoli. David Eugene Smith Collection of Historical Papers [ca. 1400–1899] (Box 3), Rare Book and Manuscript Library, Columbia University.

2. Smith Travels the World

a. Title page

b. Libri's handwritten corrections

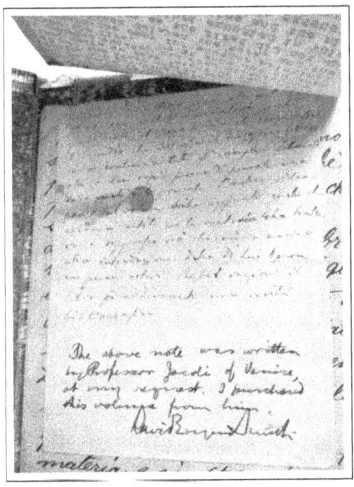
c. Note written by Jacoli, explaining the rarity of the text.

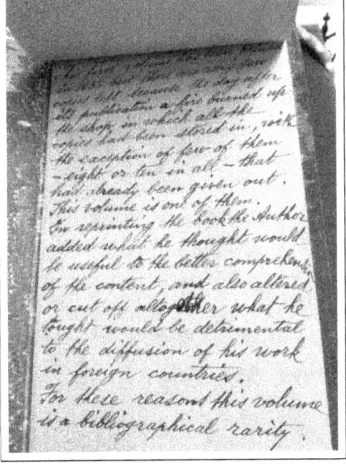
d. Note from Jacoli translated to English.

Figure 2.2.2: Libri's *History of Mathematics in Italy* (1835). The notes, among a few other pieces, are placed in the front of the text. Smith Ref R510.9 L611, Rare Book and Manuscript Library, Columbia University.

2.3 Burma, India, Sri Lanka, Japan, and China —1907

In November 1907, while on leave from Teachers College, Smith traveled to Burma and spent time in Mandalay and Rangoon. He learned that a man who owned a collection of what were said to be "interesting books" had recently died. Smith, along with his sister, his niece, and their interpreter, David Abraham, visited the man's home. Smith recollected, "I sent David up to announce our approach, and we climbed a ladder that took us up into the house. They had only one chair and that was very rickety. I was given that, [while] the ladies sat on the floor and the family... gathered in a semi-circle around me" (Smith, 1936a, p. 3). The collection consisted of a large amount of palm-leaf books related to Buddhism. Two specific items caught the attention of Smith.

> One was written on gilded copper, and the other on thick paper, also gilded. The one on copper had beautiful lettering in sepia lacquer, and the other was written in India ink. I looked at all of the books very carefully, and then I said to my interpreter: "David, there are only two books in that lot that I want. You needn't be looking at them now, and I am not looking at them, we don't want to beat the people down unduly but I want you to understand that when we leave this place you are to carry these books in your arms."... I found two long strips of linen fabric. In this had been woven the lines of a song, and while we were bargaining I asked the family if they wouldn't chant this for me, which they did. (Smith, 1936a, p. 3-4)

Smith also traveled to Lahore, India in 1907 and inquired at a local Christian college about obtaining manuscripts from dealers in the area. He learned about a Persian dealer named S. Bahadur Shah and went to his home where his collection was stored. Smith recounted the meeting as follows:

> He gave me a chair, and I told him that I wished to get any mathematical books or manuscripts that he might have. He at once said he didn't have anything of the kind, but I said "Haven't you any work on Mathematics?" and he said no, he had never bought any. He added, however, that he had a few books that I might care to look at, and so he brought out about a bushel and dumped them all on the floor. He then said that he had a few more, and would bring these in, and he left me with the pile of books. The first one I picked up was an exceedingly rare translation from the Greek into Arabic of Euclid's *Geometry*. I also found perhaps a dozen other books manifestly on Geometry or Astronomy. When he came back into the room I showed them to him and said that was what I wanted. His reply was "But I thought you wanted Mathematics!" It was evident that that term was not familiar to him. If I had said "Geometry," which is practically the same in Hindustani as in the Greek word, he would at once have known. As a result I bought, I suppose, about forty books, all of them in manuscript form. (Smith, 1936a, p. 7)

Smith reported how the dealer and he could not come to an agreement on the price for a particular manuscript written by Ulugh Beg (1394–1449). Ulugh Beg was an astronomer, mathematician, and sultan of Timurid, a Central Asian dynasty. Smith was trying to get a lower price from the dealer, and told the dealer that he was leaving the next morning to Delhi, and if the dealer would agree to Smith's price then the book could be delivered to the train station. Unfortunately, the dealer stuck to his ground, and Smith left for Delhi without the text. Smith was persistent, however, and immediately wrote to the dealer when he reached his destination and agreed to the dealer's price of $84.25. Smith explained, "I consider this the finest mathematical manuscript that I ever bought in the East. Since that time, I have purchased through him or other dealers three other manuscripts of the same work. They are older than the one that I got at Lahore, but are not so beautifully written" (Smith, 1936a, p. 8).

Although Smith was a well-known authority and respected historian of mathematics, he was not always accepted with open arms. After Lahore, Smith traveled to Bombay, where he met with a dealer of Sanskrit manuscripts. During the meeting, Smith inquired about mathematical manuscripts in Oriental languages and assured him that these manuscripts would be purchased for Columbia University for educational purposes rather than commercial sale. Smith described the experience as follows:

> [The dealer] seemed to be somewhat skeptical about this, and told me that if I would come back the next day he would consider the matter more carefully. His shop was a small room, and he must have had some thousands of manuscripts piled on the shelves there and in the adjoining room. The next day I went back to see him at the time appointed, and I found six pundits besides the dealer himself. He gave me a seat by his side, he sitting on the floor. The pundits... all sat on cushions. He then began to ask me questions as to why I wanted the books since I couldn't read them, and I told him that I was not buying them for my own direct reading, but for the reading of scholars who might be working in Columbia University in the years to come. (Smith, 1936a, p. 9)

Smith explained that the dealer wanted to know specific names of authors for the works that Smith was interested in purchasing—

almost as if the dealer was testing Smith's knowledge of Hindu writers. This was not a difficult task for Smith, and he provided the dealer with numerous names and dates. Smith was very confident in this task and believed that he knew more names than the dealer and his pundits. In fact, when the dealer and his pundits began to ask him questions on the subject, Smith turned the tables on them. He described the situation:

> I asked them about their belief in astrology, and about the difference in the works of the two Aryabhattas. They evidently had not heard that there were two, even if they had heard of one, and at the end of an hour or more they all rose and bowed to me and said that they were convinced that it was a legitimate reason that I had for wishing to get the books. [The dealer] had brought out all of the mathematical manuscripts that he had. He pointed to them and said that they would all be at my disposal. As a result, I brought out from Bombay a large number of manuscripts. They included the famous work on astronomy by Vaāhamihira. I was afterwards told that there were only six complete copies of that in Sanskrit known to Oriental scholars. (Smith, 1936a, p. 9)

This experience not withstanding, Smith did meet and interact with many people throughout his journeys who respected and admired him greatly. He visited Colombo, Sri Lanka, and visited the high priest of the Buddhists in the outskirts of the city. As Smith approached the monastery, he was greeted by three priests who told him that the high priest was not accepting any visitors at this time—due to the high priest's illness and approaching death. Smith explained further:

> I gave them my card telling my affiliation with the University and the American Mathematical Society, and

asked that it be presented to him after I left. I noticed, however, that they had called in two or three other priests, and in the corner of the room they looked over my card and then one of them left the room and I was asked to wait a few moments. Soon he returned, saying that His Reverence, the High Priest, wished to see me. I passed through a long corridor, and was admitted to his bedroom, where I saw him lying on his bed in the darkness, the curtains having been lowered. It was rather an embarrassing situation, to be going to a man's deathbed and talking mathematics to him. I found him, however, one of the finest gentlemen I ever met. He was interested in my inquiry, and he finally said that there was nothing in the library of Ceylon [Sri Lanka] that related to mathematics or astronomy. There was, however, he said, a well-known book on astrology, which necessarily had some mathematics in it, and he sent the servant out and he brought back a palm-leaf manuscript, which the high priest told me was well known among all the scholars of Ceylon. He said that it would be impossible to buy one, but that it would give him great pleasure to have a copy made and sent to me... I had been seated in a small child's chair, the priest having told me that in the presence of His Reverence no one was allowed to sit, but that in this case, if I would sit in the chair of a child, I would be welcome to that rest. When the time arrived for leaving, I arose, and to my astonishment, the dying high priest threw off the bedclothes, arose, and conducted me with great courtesy to the door. I treasure that copy that was made, not only for its own value, but also for the great kindness that this man, an invalid, showed me, a stranger. I was glad to learn afterwards that he recovered, and that he lived some years thereafter. (Smith, 1936a, p. 12-13)

During Smith's travels to Japan and China in 1907, it was his goal to purchase every worthy mathematical manuscript or book that was available. He believed that he accomplished this goal, since whenever he returned to an area, there was nothing left for him to acquire. When writing his *History of Japanese Mathematics* (1914) with Yoshio Mikami, a Japanese mathematician and historian, Smith realized that in his own collection he had all the important works needed to write the book (Smith, 1936a).

Smith could also be considered a detective in the realm of the history of mathematics. One of Smith's more famous discoveries earned him Plimpton's respect and loyalty as a fellow collector. During an early meeting with Plimpton in 1901, Smith was inspecting a manuscript of *Liber Abaci* written by Leonardo Pisano in Plimpton's collection. While comparing Plimpton's manuscript with others written by Pisano, he discovered that it was not solely the *Liber Abaci* but a compilation of 15th-century Latin manuscripts that contained a translation of part of al-Khwarizmi's *Algebra et almuchabila*. Smith loved the role of literary detective and remarked on this instance that "[t]he most interesting part of all will be to find the sources of the other extracts" (Donoghue, 1998, p. 362).

Another example of Smith as a keen investigator concerned a seal displaying the head of Galileo, which had been owned by Sir Isaac Newton, exhibited at the South Kensington Museum. Smith recognized that it was not the head of Galileo and reported this to the Museum; it was immediately removed from exhibition, despite having been owned by Newton (Smith, 1936a).

During his travels to China in 1907, Smith hurriedly purchased several volumes on geometry. Unfortunately, he did not have any time to translate the texts, but did note they were written by Li

Ma Do. Smith recollected:

> I never had heard of such a mathematician, but I bought the book and shipped it with the other material to New York. Coming up the Red Sea, one day, I kept thinking about this book of Li Ma Do's, and then suddenly it came to my mind who the man was. I knew that the Chinese could not pronounce 'r' and used 'l' instead... then the name of the author might have been Ri Ma Do. Then it suddenly dawned upon me that the principal name, which we would put at the end of a signature, ought to go at the beginning, according to their plan. The author, then, might have been Madori, and then the whole thing came to me. 'Mado' was 'Matteo', the 'Ri' was the beginning of 'Ricci', and I apparently had a manuscript copy of the translation of Matteo Ricci from a Latin edition of Euclid into Chinese. I have now in this library a complete manuscript of this epoch-making translation of Euclid, the first that was attempted in the Far East. (Smith, 1936a, p. 16–17)

Throughout this particular journey, Smith collected the majority of his items of Japanese, Chinese, and Indian origin (Broomell, 1908). Although Smith was always on the lookout for curios, he was a dedicated professor at Teachers College until his retirement in 1926. Afterwards Smith continued his devotion to expanding his collection while traveling around the world. Some trips abroad had devastating implications for Smith and his family. During a trip to Paris in July 1922, Smith's wife, Fannie, suffered injuries in a car accident that hampered her health for the rest of her life. She passed away in 1928—just before her 63rd birthday (Smith, 1922). Smith waited twelve years before his second marriage, at age 80, to fellow educator Eva May Luse in 1940.

2.4 Sumatra, Thailand —1930

His acquisitions often took Smith to unique locations, where he met curious people of all ages. In 1930, Smith traveled to Sumatra and visited the Batak Museum. The curator told him that there was only one book related to Smith's interest in a small village not far from where they were. Smith remembered:

> We [Smith, his sister, and niece] climbed over the mud walls surrounding the little village, and asked to see the headman. He came out dressed in a scanty robe and one tooth hanging in the front of his mouth. I stated my problem to him, and he said, yes, they had a copy of a book of that kind, and he sent out to get it. Meanwhile [the entire] village had gathered around us—men, women, children, poultry and pigs... I asked him to read some of [the book] for me, which he did. This being [said] to the chauffeur and thence to the interpreter and thence to me. It seems that the book has a little astrological material, but chiefly it was made up of incantations of one kind and another. The headman told me that there were a number of medical recipes in it, but that it was very potent in relieving illness by being bound on the part of the body that had the most pain. If that were done, the pain would all be gone the next day. This piece, therefore, has probably been bound upon hundreds of sores and ailments of every kind. Not leprosy, however! He also said that they had some writing on bamboo sticks, which contained charms against disease... The writing is scratched with a jackknife and rubbed in with lamp black. After a good deal of bargaining, I bought two of the bamboo cylinders and the one book, which was said to be the only one on the is-

land outside of the museum. We then climbed over the mud wall and walked to... our car. (Smith, 1936a, p. 1-2)

People with items to sell also sought out Smith, including friends, colleagues, scholars, and even young children. As described by Smith following the meeting with the headman that day:

[W]e were going up the hill, [and] I noticed a boy running across the meadow and waving his hands at us... [he] stood in the road so that we couldn't pass. He came to the automobile with two other books in his hands. He said we must not tell the headman of this, but that they had had two books in their family and he wanted to sell them. Then resulted the usual bargaining, and I came away with the books, so that instead of there being only one known book on the island there were at least three, and I have no doubt there were others. (Smith, 1936a, p. 2-3)

During Smith's 1930 trip to Siam he visited the museum in Bangkok. The curator of the museum told Smith that in order to purchase any materials from dealers in the area it would have to be cleared by Prince Damrong Rajanubhab (1862–1943). Prince Damrong was the founder of the modern Thai education system and an avid historian (Bunnag, 1977). Following directions of the curator, Smith wrote a letter to the Prince noting that he was interested in obtaining any items related to mathematics or astrology "that showed some excellence in calligraphy" (Smith, 1936a, p. 18). This message was immediately sent over to the Prince, as Smith received a response only ten minutes later from him asking Smith to come visit at once. Smith described the visit as follows:

> My sister and niece were with me and I couldn't take them around there because they had not been invited, but we took our carriage, automobiles being somewhat out of the question at that time, and we drove around to the entrance to the grounds and were at once admitted. I told the driver to take me to the palace and then take my sister and niece up to a place I saw in the park where there was a good shade. Soon, however, we reached the palace. The door opened, and the Prince came out, and I apologized for bringing my sister and niece when they had not been invited to come. He was most gracious, and at once invited all three of us to have tea with him. Then began an acquaintance, which lasted for two or three years. The result of the whole matter was that he gave me the name of the only dealer of any consequence in the town, and gave me much information concerning the kind of books that would be available. As a result of all that, there were two barrels of manuscripts... to be sent to America. I was entertained the next day at a tea given by the librarian [at the palace] and I met there a number of the literati. I said to the Prince that I understood that I could not export any books or manuscripts without the permission of the Government, and that that would take a couple of weeks, whereas I had to leave in a few days. He smiled pleasantly, and said I should leave that to him. The books would be examined quickly, and there would be no question about the transportation. (Smith, 1936a, p. 18)

Smith and his family met the Princess, daughter of Prince Damrong, during their visits to the palace and remained in contact with both the Prince and Princess until a coup d'état in 1932 when the family was exiled to Malaysia (Smith, 1936a).

 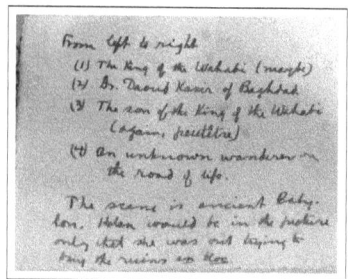

a. Smith sent this photograph to Miss Bertha M. Frick during his travels through Persia in 1933.

b. The reverse side of the photograph where Smith demonstrates his humorous side with witty notes, such as describing himself as "an unknown wanderer on the road of life."

Figure 2.5.1: David Eugene Smith Professional Papers, 1860–1945 (Box 48), Rare Book and Manuscript Library, Columbia University.

2.5 Persia —1933

In 1933, Smith published his translation of the *Rubaiyat* of Omar Khayyam. As a consequence of receiving decoration from the Shah of Iran (Fite, 1945) for the translation, he became popular as a potential buyer of antiquities in Persia (Figure 2.5.1); in fact, the *literati* of the area gave a reception in his honor when he arrived. Many dealers who had searched for and obtained numerous books on Persian poetry and mathematics for Smith to purchase approached him, and Smith purchased about twenty manuscripts. In this group was a manuscript from the tenth century, a translation from Greek into Arabic of Archimedes' work on the sphere and cylinder (Smith, 1936a).

As can be expected in traveling throughout the world, there were times when Smith was not completely safe. One specific incident occurred in Teheran, the capital of Iran, during the month when pilgrimages were made to Meshad. Smith was on a mission to view a specific Koran that was located in a shrine in the city. Since this was during the holy week, it would be sacrilegious for

Smith to enter the shrine, as he was an infidel. The Governor, who was considered an open-minded man, told Smith he would do everything that he could to let him see what was considered the most beautiful example of calligraphy in existence. He warned Smith, "I can get you into the shrine, but I can't get you out alive." However, he said that he would find a way to get him into the shrine and get the book out for Smith to view (Smith, 1936a, p. 22). Smith retold the event:

> At the prescribed hour I went there, leaving the ladies [Smith's sister and niece] in the compound, for they of course were not permitted anywhere near the holy place. They in fact apparently knew nothing about the projected adventure. The Chief of Police, the Secretary of the Governor, the Secretary of Education, and one other person, took me around through one of the bazaars to a side entrance to the compound. I was dressed with the Persian hat, but with European clothes—which was not at all uncommon in the case of Mohammedans. I was taken in as an Egyptian who lived in Istanbul and couldn't speak Persian, so they had to translate for me. Well, we walked in. I was in the middle, and I was warned not to speak a single word. As a matter of fact, I got enthusiastic and did speak one word. At that the Chief of Police squeezed me so hard in warning that he almost broke my hand. He was afraid someone might hear me. I passed as far away as possible from three or four priests who were preaching to groups of the Faithful, all sitting on the ground. I saw one of the priests looking at me in a very curious way, and I looked at the ground—in a very curious way! We spent about a half an hour going around, and finally reached the entrance gate to the compound, and I must say I drew a

long breath of relief when I got out. I returned to the Governor's house, and he said, 'I will get the book you want to see.' He did more than that, though. He sent out to the librarian of the shrine, that was the Holy of Holies, and they brought as many as ten large manuscripts and put them on the table in front of me. For two hours I sat there and talked to the Governor and held in my hand that one book, which I still think is the most beautiful book in the world... They had brought over the best books they had to show me—this, too, seems to have been strictly against the Mohammedan tenets, for they were not supposed to be looked at by an infidel—but apparently the librarian of the shrine was another 'broad-minded' man. But they were taking no chances of losing them and saw to it that they were well guarded. The head librarian came himself, and brought his servants as well. Besides this they carried the books over to the Governor's house under a cover so that no one would know that they had been taken out. (Smith, 1936a, p. 23-24)

These excerpts illustrate Smith's spirit of adventure and his zeal in pursuit of materials related to the history of mathematics. Over years of international travel by steam ship, train, bus, car, buggy, and cart Smith pursued the books and artifacts that made up his collection. He seems to have been ready to undertake any journey and considerable personal risk to add a new treasure to his trove.

Travel, however, was only one aspect of Smith's life. After joining the faculty of Teachers College in 1901, Smith's energies also focused on the organization and display of his growing collection and its value for bringing the teaching and learning of mathematics to life from his New York home base. The next chapter presents an in-depth look at Smith's influence on the initial design of the Educational Museum, beginning with a brief overview of the origins of Teachers College in the 1880s through the merger of the college with Columbia University in 1898.

2.6 References

Broomell, B. L (1908). Mathematical historical material from the east. *Mathematics Teacher*, *1*(1), 14-15.

Bunnag, T. (1977). *The provincial administration of Siam, 1892-1915: the Ministry of the Interior under Prince Damrong Rajanubhab.* Kuala Lumpur; New York: Oxford University Press.

Dauben, J. W. (2002). David Eugene Smith. In J. W. Dauben & C. J. Scriba (Eds.), *Writing the history of mathematics: Its historical development* (pp. 524-525). Boston: Birkhauser.

Donoghue, E. F. (1998). In search of mathematical treasures: David Eugene Smith and George Arthur Plimpton. *Historia Mathematica*, *25*(4), 359-365.

Fite, W. B. (1945). [Obituary]: David Eugene Smith. *The American Mathematical Monthly*, *52*(5), 237-238.

Ginsburg, J. (1944, September 21). [Letter to Bertha M. Frick]. David Eugene Smith Professional Papers, 1860-1945 (Box 19). Rare Book and Manuscript Library, Columbia University, New York.

Lawler, T. B. (1938). *Seventy years of textbook publishing: A history of Ginn and Company.* New York: Ginn and Company.

Murray, D. R. (2012b). David Eugene Smith's adventures in collecting. *International Journal for the History of Mathematics Education*, *7*(1), pp. 51-60.

Russell, J. E. (1937). *Founding Teachers College: Reminiscences of the dean emeritus.* New York: Teachers College, Columbia University.

Simons, L. G. (1945). David Eugene Smith—in memoriam. *Bulletin of the American Mathematical Society, 51*, 40-50.

Smith, D. E. (n.d.a) David Eugene Smith: A sketch. David Eugene Smith Professional Papers, 1860-1945 (Box 64). Rare Book and Manuscript Library, Columbia University, New York.

Smith, D. E. (1922, July 5). [Letter to George A. Plimpton]. George Arthur Plimpton Papers, 1650-1956 (1877-1936) (Box 7). Rare Book and Manuscript Library, Columbia University, New York.

Smith, D. E. (1936a). [Unpublished little folding book]. David Eugene Smith Professional Papers, 1860-1945 (Box 115). Rare Book and Manuscript Library, Columbia University, New York.

Smith, D. E., & Mikami, Y. (1914). *A History of Japanese mathematics.* Leipzig: F. Meiner.

Chapter 3

Early History of Teachers College and the Educational Museum

The history of Teachers College starts in 1880 when the Kitchen Garden Association was established in New York City by Grace Hoadley Dodge (1856–1914) (Figure 3.0.1). The Kitchen Garden Association was a philanthropic institution for young girls to learn the domestic industrial arts (Kitchen Garden Association, 1880).

It was soon realized that the "teachers" of these students were in need of their own proper training, which resulted in official instruction classes taught by the leaders of the Association. After four years, the leaders of the Kitchen Garden Association realized that the association grown both in enrollment and philosophy and thus was no longer simply a philanthropic institution but had become an educational institution.

3.1 Industrial Education Association

In 1884 the Kitchen Garden was reorganized as the Industrial Education Association. This new association allowed both boys and adults to enroll, held more sophisticated classes in the industrial

3. Early History of the Educational Museum

Figure 3.0.1: Grace Hoadley Dodge (1856–1914). Image provided courtesy of the Gottesman Libraries at Teachers College, Columbia University.

arts, and included both men and women as organizational leaders. The president of the Association was General Alexander Webb, who was also the president of the College of the City of New York at the time, with Grace Dodge as the vice-president. Since Webb was preoccupied with other work commitments, Dodge would assume presidential roles on many occasions (Cremin, Shannon, & Townsend, 1954).

The Industrial Education Association made an effort to publicize its focus on education by organizing the Children's Industrial Exhibition in April of 1886. James Earl Russell, future dean of Teachers College, described the exhibition in his *Founding of Teachers College* (1937), as "participated in by some sixty schools and institutions, and attracted visitors from many other cities and led eventually to the introduction of household arts and manual training into the schools of New York City" (Russell, 1937, p. 5). The items that were on display in this exhibition eventually became part of what would be known as the Educational Museum of Teachers College.

3. Early History of the Educational Museum

By the end of 1886, the Industrial Education Association had outgrown its original location and moved to the building where the Union Theological Seminary was located at No. 9 University Place. This building was described as, "a five-story structure... [containing] offices, an assembly hall, and a model cooking room on the first floor, classrooms and a large museum for exhibits on the second. The three top floors were reserved for residents, while in the basement was located a 'little housekeepers' classroom" (Cremin et al., 1954, p. 13–14). The Association had minimal funds to lease such a building, but Dodge stepped in and paid for the space from her own finances for eight and a half years. The official opening of the new building was on December 14, 1886, where the President of John Hopkins University, Daniel C. Gilman, was the main speaker at the opening ceremony (Cremin et al., 1954).

The popularity of the Children's Industrial Exhibition led to a decision that the Association needed to focus on education, and in particular become a facility for the training of teachers. Expanding enrollments and a lack of properly trained teachers created in turn the need for a Training College. This aspect of the Association demanded that an experienced educator to be at its head. In February 1887, Professor Nicholas Murray Butler (1862–1947) (Figure 3.1.1), from Columbia College's Department of Philosophy, became the president of the Industrial Education Association. One of Butler's major contributions to the Industrial Education Association was transforming it into the New York College for the Training of Teachers. Butler kept the "Number 9" model school intact, though he renamed it The Horace Mann School (Cremin et al., 1954). Horace Mann (1796–1859) was the first Secretary of the Massachusetts Board of Education, as well as a member of the United States House of Representatives. He was a revolutionary in the field of education and believed that every person should receive public education. This school still exists today (Horace Mann School, 2010). Thus Butler and his administration laid the groundwork for the educational mission of what would eventually become Teachers College.

Figure 3.1.1: Nicholas Butler (1862–1947). Image provided courtesy of the Gottesman Libraries at Teachers College, Columbia University.

The idea of teacher education as the focus of a school was not new to Butler. During his time at Columbia College, he became close friends with its president, Frederick Barnard (1809–1889). Barnard was passionate about the importance of the study of education. In fact many years prior, in 1858, Barnard had proposed that a school of education be established at the University of the South. He made a similar proposal in his inaugural address at Columbia College in 1881, where he wished to create a department for the study of education. The Board of Trustees at Columbia College, however, did not share this vision and vetoed the establishment of such a department.

Unwilling to give up on this goal, Barnard enlisted his student, Butler, to join forces with him to incorporate the study of education into Columbia. After Butler graduated with his Ph.D in 1884, he returned to teach at Columbia and assist Barnard in convincing the Trustees that an education department would be a significant and important addition to the College. In 1886, Barnard and Butler organized a series of lectures on the topic, which they believed would motivate the public and persuade the Board of Trustees. They were proven correct in one aspect: the lectures were "attended by more than two thousand hearers" (Russell, 1937, p. 6). However, the strategy failed regarding their major objective. The Board of Trustees at Columbia was still skeptical and rejected the proposal yet again.

3.2 Teachers College

Butler's acceptance of the presidency of the Industrial Education Association in 1887, and his role in founding the New York College for the Training of Teachers provided another avenue for Butler and Barnard to demonstrate that the study of education and the training of teachers was a valuable and necessary service for institutions of higher education (Cremin et al., 1954). During his presidency, Butler promoted his views on education by initiating multiple publications such as the *Educational Leaflets*, the *Educational Monographs*, the *Educational Review*, and the "Great Educators Series." Through his efforts, the Training College gained a respectable reputation throughout the country. Walter L. Hervey (1862–1952), the Dean of the Training College, took over the presidency in 1891 when Butler was promoted to the head of the Department of Philosophy, Ethics, and Psychology at Columbia College. In December 1892, the name of the New York College for the Training of Teachers was officially changed to Teachers College (Cremin et al., 1954).

After Butler's resignation as president, Teachers College lost its momentum for independent growth and the school's leadership looked once again to Columbia. In 1892, the Trustees of Teachers College voted that the best course of action would be to hand over administration of Teachers College to Columbia College. This was a one-sided decision, however. As before, the Board of Trustees at Columbia College refused the proposal to include a department of educational studies under its name, stating, "there is no such subject as Education and moreover it would bring into the University women who are not wanted" (Russell, 1937, p. 26). They did agree upon a somewhat limited alliance in 1893, to the effect that Columbia would use Teachers College as a place to send students for pedagogy instruction and Teachers College students and faculty would be allowed to avail themselves of Columbia University's instruction, scholarship, atmosphere, and library (Cremin et al., 1954).

Despite these struggles at the administrative level, the enrollment at Teachers College and The Horace Mann School continued to grow, outpacing the capacity of No. 9 University Place (Figure 3.2.1). A location in upper Manhattan seemed the perfect spot for the relocation of the expanding institution. Due to last minute donations by George W. Vanderbilt (1862–1914) and an anonymous donor from New Zealand, Teachers College found its current home on the north side of 120th Street (Russell, 1937). At the same time, Columbia College expanded and purchased the land on the south side of 120th street. It was not by coincidence that these two colleges ended up in such close proximity. Butler's vision of bringing the institutions together was still at work. Dr. Butler was a member of Columbia's relocation committee and asked Vanderbilt, a Trustee of Teachers College, to choose a location adjoining Columbia (Cremin et al., 1954). Teachers College officially moved into Main Hall on 120th Street in 1894.

Another major figure in the history of Teachers College is James Earl Russell (1898–1926) (Figure 3.2.2), Dean of the Col-

3. Early History of the Educational Museum

Figure 3.2.1: Arithmetic Class. Seventh Grade. No. 9 University Place, Teachers College, 1893. Image is provided courtesy of the Gottesman Libraries at Teachers College, Columbia University.

lege from 1897–1927. Russell, who originally came to Teachers College as a professor in the technical training of teachers, quickly became dean due to his educational viewpoint, the unsettled nature of Teachers College, and the influence of President Seth Low of Columbia (Cremin et al., 1954). At the time, for Teachers College to survive as an institution, it seemed necessary to become a professional school of education under a university, and Columbia was the obvious choice. Russell provided the basis for yet another proposal for integration and President Low, who also had an appreciation for the study of education, gave his support to the plan. This new proposal met with success: the Board of Trustees of Columbia finally agreed to a partnership in 1898.

Figure 3.2.2: James Earl Russell (1898–1926). Image provided courtesy of the Gottesman Libraries at Teachers College, Columbia University.

3.3 The Educational Museum

Now that Teachers College was part of Columbia, Dean Russell continued to pursue his vision for the broader educational mission of the school, including the expansion of its original collection of educational items gathered for the Children's Industrial Exhibition in 1886. The Industrial Education Association, housed at No. 9 University Place, had many exhibits scattered throughout the school's departments of manual training, art, domestic science, domestic art, and natural science—with no real "home" (Andrews, 1909). The Exhibition's contents continued to be decentralized and dispersed among various departments after its relocation to 120th Street.

Dean Russell officially established the Educational Museum in 1899 with the appointment of a curator, George S. Kellogg (Russell, 1899). Russell believed in the importance of the Museum and thought it would serve "to provide a systematic way of collecting

3. Early History of the Educational Museum

illustrative material" (Russell, 1899, p. 16). Russell's vision required a dedicated space for the Museum and its collection. In 1901, the Educational Museum was provided with a space of its own on the second floor of Main Hall—currently Zankel Hall of Teachers College—in room 215 (Figure 3.3.1). The Educational Museum grew quickly during the next ten years to become a well-regarded showcase of educational objects and historical artifacts of many subject areas, including mathematics (Andrews, 1909).

Figure 3.3.1: Educational Museum of Teachers College. Image is provided courtesy of the Gottesman Libraries at Teachers College, Columbia University.

Teachers College was not the first or the only institution intent on building such a museum collection to support teaching, school curricula, and the study of science and practical arts. In his 1909 study of the history of museums of education, Andrews defines this type of museum as "an institution that contains objective collections which have an illustrative, comparative, or critical relation to the schools and to school work, or which are concerned with education as a profession, a science, or a social institution" (Andrews, 1909, p. 3). Andrews reports that the first educational museums were founded in Europe during the 1860s and that the trend had continued and expanded throughout the world up to the time of his study. At the beginning of the twentieth century, he noted that a renewed interest in educational museums occurred in the United States with a number of cities and American universities taking steps to create their own version of an educational museum for the use of students and the public. St. Louis, for example, had a major educational museum that, with some modifications, is still in existence today. The University of California at Berkeley, Clark University, Harvard University, the University of Illinois at Champaign, and Indiana University at Bloomington had all established educational museums by 1909. The collections in each of these institutions varied in content and organization. For example, Harvard did not maintain an educational museum on its campus but rather exhibited local students' work in towns throughout Massachusetts. Indiana University had a centralized collection and used its material as a type of reference book for the teachers, similar to that of a library. Andrews notes that at the time of his study, in addition to the popularity of the American educational museum movement, there were also seventy-four educational museums outside of the United States. His 1909 study organizes all these in one directory (Andrews, 1909).

The creation of the Educational Museum in 1899 was thus part and parcel of a broader worldwide educational trend. At this time the mission and scope of Teachers College had expanded to include two kindergartens, two elementary schools, a high school, an undergraduate college, and a graduate school. This span of educational activity made the creation of a larger museum to support teacher training and curriculum a natural match.

Commenting on the museum's founding, the 1899 *Columbia University Quarterly* noted:

> A unique feature of the report is the discussion of the possibilities and the requirements of an educational museum, designed to illustrate various educational systems, class work and general educational problems and meth-

3. Early History of the Educational Museum   43

ods. Teachers College has definitely taken up this museum problem, and a good beginning has been made toward making available in all departments the photographs, lantern slides and other illustrative material scattered about the buildings. (p. 80)

Within two years of its opening, the Educational Museum was gaining popularity due to its special exhibits in various fields, including anthropology, oriental art and industry, textbooks, and religion (*Columbia University Quarterly*, 1902). The Museum was poised for growth in its collection as well as its influence on the teaching profession. Smith arrived on the scene ready to provide leadership in both areas.

Smith joined Teachers College in 1901 after establishing an international reputation as a leading teacher, author, and collector in the history of mathematics. His arrival at Teachers College during this formative period gave Smith the opportunity to influence the growth and organization of the Educational Museum, especially its mathematical collections. He brought an impressive background to the task. In 1902, Smith had been the chairman of the educational exhibit at the St. Louis Exposition, the precursor to the St. Louis World Fair in 1904. This exhibition had been awarded a gold medal.

Recognizing his expertise, Russell asked for Smith's advice on how the Educational Museum could be better organized. In an eleven-page letter to Russell on December 8, 1902, Smith stated: "You have twice suggested that I tell you definitely what I think should be done with the Educational Museum. My proximity to the room has led me to see more of it than the rest of the faculty, and a rather natural instinct for collecting has stimulated my interest in it, and therefore I am prompted to follow your suggestion" (Smith, 1902, p. 1).

Smith had a specific vision for what the new Educational Museum could become with the proper guidance. He began the letter by stating that the museum should not be merely a storeroom for materials needed by the faculty. Smith believed that this type of material should be kept in each department, for easy access. He further explained that the museum should not solicit its material from dealers, which would cause unnecessary items being included in the collection. A final, but major point, Smith stressed, was that the current curator, George Kellogg, should be removed from the position. Smith was adamant about this, commenting: "I find that the faculty will positively decline to take any interest in [the museum] so long as he has control of it" (Smith, 1902, p. 2).

In the letter, Smith went on to make very specific suggestions, down to what should be in the cases and how the cases should be organized. He proposed that there should be "a series of large cases, one for each grade from kindergarten through the high school... [it should include] the world's best illustrative material... [yet avoid] the extravagant, the bizarre, and all that is not usable and valuable for our American schools" (Smith, 1902, p. 3). Smith thought of the museum as a place for teachers to visit and view not only the historical side of education but also its practical side: "a label on each piece of material should state its purpose, the price, and where it can be purchased. A superintendent, principal, or teacher should be able to go to the cases and see the best and most modern material, and ascertain the prices and the makers, without looking over catalogues or making further inquiry" (Smith, 1902, p. 3).

As can be expected, Smith valued instruments and tools in education, hence he wanted to have these types of items properly showcased in the museum. He specified the actual apparatus to be displayed rather than the student's work with the tool. Smith had an idea for the subject organization as well: "handwork (only the tools, not the product), domestic art, domestic science, fine art, geography, mathematics, [and] natural science" (Smith, 1902, p. 4).

Smith did value student work, however, especially that from other countries. He told Dean Russell that "after such a collection has been established, and possibly while establishing it, the best specimens of pupils' work should be collected, and this, too, should be arranged by grades" (Smith, 1902, p. 4). He discussed further that Teachers College already had some examples of work from Japanese students; however, since it was not properly organized and far from being complete, it had "no real value" (Smith, 1902, p. 5). All of these suggestions were focused on the practical part of the museum, as opposed to its historical items. In Smith's final suggestion for what the museum could become he stated, "not a thing should go into the collection of school supplies that we cannot indorse, and when it becomes obsolete or is considered unworthy, it should be thrown out entirely, or put in a case of historical material" (Smith, 1902, p. 5).

In these early years, Smith had high expectations for the Educational Museum. He had visited a few other similar organizations; however, in Smith's opinion, there were remarkably few successful models despite the large number of museums in Andrews' catalog. According to Smith, there existed only three or four "serious attempts" at model educational museums, and in his opinion they were failures due primarily to a lack of organization. Citing examples of failed attempts in Russia, Germany, and France, as well as his own attempts in Michigan, Smith noted, "the movement was started at Ypsilanti five years ago [1897], but on a wrong basis, and it failed. It is, I believe, safe to say that there is, today, no well arranged, modern, helpful collection of this kind anywhere" (Smith, 1902, p. 7). Nonetheless Smith was optimistic that with the proper guidance and adequate support from the administration, the Educational Museum could be a pioneer and an example to all in the field. He estimated that the cost of establishing such a museum would be $2000, not including the cost of the cases. He further suggested to Russell that $500 a year would keep it running, while $1000 a year would be needed for all other necessities besides wages (Smith, 1902).

Smith concludes his letter by drilling in the point to Russell that the success of the Museum was contingent upon Kellogg being replaced by a faculty member acting as curator. This would ensure the cooperation from the faculty, which, it seems, was lacking with Kellogg as curator. Smith understood that it would be a huge undertaking for a faculty member and suggested to Russell to allow for additional remuneration. He suggested two colleagues for the position, Professors Samuel Dutton and Frank McMurry (Smith, 1902, p. 11). Dutton was the Superintendent of the Teachers College Schools and a Professor of School Administration and McMurry was a Professor of the Theory and Practice of Teaching (*Columbia University Catalogue*, 1901).

Even though at this moment Smith was relatively new to Teachers College, Russell acknowledged that Smith was an authority on these matters. That Russell followed many of Smith's recommendations is evident from Andrews' 1909 description of the organization and management of the Educational Museum of Teachers College. During the 1907–1908 academic year, Andrews reports that there were many specified collections: curriculum and methods of elementary and secondary education, educational administration, school buildings and equipment, history of education, foreign school systems, art, biology and nature study, domestic art, domestic science, geography, history, kindergarten education, language and literature, mathematics, manual training and industrial arts, natural science, physical education and anatomy, and religious education (Andrews, 1909). Although extended, this list was very similar to Smith's in 1902.

3. Early History of the Educational Museum

Russell took Smith's advice and within a year Kellogg had been removed from his position as curator. In 1903 Benjamin R. Andrews was appointed Supervisor of the museum. He remained in that position until 1906. David S. Snedden, Adjunct Professor of Educational Administration, followed Andrews as Director (Andrews, 1909). In 1909, David Eugene Smith took over supervision of the museum as curator and director (*Columbia University Quarterly*, 1909, p. 97).

During the Museum's active years, there were exhibits in varying disciplines. One that is similar to Smith's collection was an exhibit of George Plimpton's mathematical books. In a letter to Smith dated December 4, 1902, Plimpton states, "next winter I want to exhibit my mathematical books and get you to give a talk on them" (Plimpton, 1902, p. 1). This exhibit occurred during the fall semester of 1903. Andrews wrote to Plimpton on December 11, 1903 and described how over 1,700 people visited the exhibit during the two weeks. He continued "Dr. Smith has doubtless already told you of the value of this particular exhibit to his students and the College at large, as well as to the many outsiders who came in" (Andrews, 1903, p. 1).

In Andrews' 1909 study, the Educational Museum is described as a well-organized system, possibly due to Smith's urgency to the Dean in 1902. They used the "Dewey numerical classification" for the lantern slides and photographs, along with card catalogs for the collections as a whole. There were case guide cards to provide even more clarification for a direct reference to where the item would be located in the Museum. The care taken in designing the display cases was recognized by many museums, so much so that they copied them for their own collections (Andrews, 1909, p. 27).

The Educational Museum used special exhibits to present the materials in Smith's collection both in the designated museum

3. Early History of the Educational Museum

Figure 3.3.2: Photograph of the Department of Mathematics exhibit room, circa 1903. Image is provided courtesy of the Gottesman Libraries at Teachers College, Columbia University.

rooms and in the Department of Mathematics offices. As early as 1904, Smith was opening up his collection to students, as noted in the 1904–1905 *Columbia University Quarterly*, "Professor Smith, of the department of mathematics, has made available for the use of students his private mathematical library of 4,000 volumes and 6,000 pamphlets, and his unique collection of 2,000 manuscripts and 1,100 portraits of mathematicians. Selections from the last have been reproduced for use in other institutions" (*Columbia University Quarterly*, 1905, p. 379–380).

The Department of Mathematics of Teachers College had its own Mathematical Library in Room 212 Main Hall. This was where much of Smith's large personal collection of books, pamphlets, instruments, manuscripts, engravings, and portrait medals were displayed (Figure 3.3.2). The adjoining room, Room 211,

3. Early History of the Educational Museum 49

contained the collection of mathematical apparatus and models related to number games and mensuration (Department of Mathematics of Teachers College, n.d.; Figure 3.3.3). Smith's collection in the Department of Mathematics was actively promoted through detailed descriptions in journal articles and bulletins. For example, a 1907 piece in *Science* was entitled, "A Mathematical Exhibit of Interest to Teachers" (A Mathematical Exhibit, 1907).

The next chapter presents an in-depth look at the highlights of the mathematical collection as it was viewed by Smith and his contemporaries in the first decade of the twentieth century, during what turned out to be a high point of public and institutional interest in the Museum's mathematics materials.

3. Early History of the Educational Museum

a. Facing towards the door

b. Facing opposite direction

Figure 3.3.3: Photographs of the Department of Mathematics exhibit room. David Eugene Smith Professional Papers, 1860--1945 (Box 121), Rare Book and Manuscript Library, Columbia University.

3.4 References

A mathematical exhibit of interest to teachers. (1907). *Science, XXV*(632), 232–234.

Andrews, B. R. (1903, December 11). [Letter to George A. Plimpton]. George A. Plimpton Papers, 1634–1956, (Box 46). Rare Book and Manuscript Library, Columbia University, New York.

Andrews, B. R. (1909). *Museums of education: Their history and their use* (Doctoral dissertation). Teachers College, Columbia University, New York.

Columbia University Catalogue. (1901). Teachers College. Columbia University: New York

Columbia University Quarterly. (1899). Teachers College. Columbia University: New York.

Columbia University Quarterly. (1902). Teachers College. Columbia University: New York.

Columbia University Quarterly. (1904). Teachers College. Columbia University: New York.

Columbia University Quarterly. (1909). Teachers College. Columbia University: New York.

Cremin, L. A., Shannon, D. A., & Townsend, M. E. (1954). *A history of Teachers College, Columbia University.* New York: Columbia University Press.

Department of Mathematics of Teachers College. (n.d.) [Pamphlet describing David Eugene Smith's Collection]. David Eugene Smith Professional Papers, 1860-1945 (Box 137). Rare Book and Manuscript Library, Columbia University, New York.

Horace Mann School. (2010). A long tradition.

Kitchen garden association: What is being done throughout the United States to increase the supply of well-taught house servants. (1880, November). *The New York Times.*

Plimpton, G. A. (1902, December 4). [Letter to David Eugene Smith]. David Eugene Smith Professional Papers, 1860-1945 (Box 39). Rare Book and Manuscript Library, Columbia University, New York.

Russell, J. E. (1899). Report of the Dean. *Teachers College Bulletin.* New York: Teachers College.

Russell, J. E. (1937). *Founding Teachers College: Reminiscences of the dean emeritus.* New York: Teachers College, Columbia University.

Smith, D. E. (1902, December 8). [Letter to James E. Russell]. David Eugene Smith Professional Papers, 1860-1945 (Box 41). Rare Book and Manuscript Library, Columbia University, New York.

Chapter 4

Step Inside Smith's Collection in the Educational Museum

4.1 D. E. Smith's Collection in His Own Words

The years around 1909 seem to be the pinnacle for the Educational Museum as well as for interest in Smith's collection at Teachers College. This could be due to the fact that 1909 was the year when Smith, who had long been an enthusiastic proponent of the museum, officially took over responsibilities as the Director of the Museum. During July of that year, the entire collection was recognized in *Scientific American* in a multi-page article containing photographs of the remarkable instruments in Smith's collection (Wade, 1909). At this time Smith, through the Department of Mathematics, printed "several facsimile pages from a manuscript of about 1300 A.D., representing the earliest English use of Arabic numerals, and has provided for the use of students an illustrated catalogue of Professor David Eugene Smith's collection of one hundred and twenty-two portraits of Sir Isaac Newton" (*Columbia University Quarterly*, 1909, p. 97). Smith's mission was always to allow as much access to his collection as possible. The Educational Museum also loaned its material; in 1910, for example, it made about 11,460 loans, one-fifth of them to other institutions (*Columbia University Quarterly*, 1910, p. 325).

The 1910 "Bulletin of the Buffalo Society of Natural Sciences" was dedicated to producing a Directory of American Museums. The Educational Museum was included in this directory, and Smith is named as its curator with Sarah Mitchell Neilson as assistant. It describes the Museum's three functions: "a repository of exhibits showing the work of various departments... an agency to collect and circulate illustrative material for the use of the college and its schools... a place for temporary exhibits of educational nature, about [six] of these being held during the academic year" (Buffalo Society of Natural Sciences, 1910, p. 203). It also states that the museum was completely funded through the general budget of Teachers College (Buffalo Society of Natural Sciences, 1910).

As described in the previous chapter, the Teachers College Department of Mathematics also had an extensive collection of artifacts, historical documents, portraits, and mathematical tools housed in Rooms 211 and 212 in Main Hall (Figure 4.1.1). This collection of Smith's items became very well known in other schools and colleges—in fact the exhibits at Teachers College were open for student excursions to permit students to view and use the instruments on site (Peters, 1911). Smith knew that his collection was quite unique and valuable. In a letter dated April 12, 1911 addressed to Dr. C. T. McFarlane, the Controller of Teachers College, Smith made the argument that Teachers College should provide insurance for his collection. At that time, Smith valued his collection at $15,000 (Smith, 1911).

Smith became deeply involved with the exhibition of his own collection, as well as with all aspects of the organization and management of the Educational Museum. He gave lectures about the

4. Step Inside Smith's Collection

Figure 4.1.1: Photograph of the Department of Mathematics exhibit room, circa 1903. Image is provided courtesy of the Gottesman Libraries at Teachers College, Columbia University.

mathematical materials, organized tours for those who could visit in person and responded to many requests for information from around the world. The educational impact of Smith's collection and his efforts to make it available far and wide are difficult to grasp from our current perspective of instant access to digital information. The goal of this chapter is to provide a window into what it was like for students and teachers to visit the Educational Museum in its heyday, to walk through rooms filled with rare mathematical artifacts, to attend one of Smith's talks on the objects, and to sense the curiosity from the public regarding the collection.

A major source document for this inside tour of the collection is a talk given by Smith for the Hobby Club of New York City on January 25, 1917. The Hobby Club's mission was "to encourage the collection of literary, artistic and scientific works; to aid in the development of literary, artistic and scientific matters; to promote social and literary intercourse among its members, and the discussion and consideration of various literary and economic subjects" (Hobby Club, 1920, p. 9). George Plimpton was elected to the Hobby Club in 1911, and Smith followed in 1913. As the name implies, each member of this club had a distinct type of hobby and collection, such as rare books, clocks, Shakespeare, or mathematics. Smith's notes for his talk, titled "Mirabilia Mathematica," are included in the David Eugene Smith Professional Papers collection in the Rare Book and Manuscript Library at Columbia University. The following sections describe the highlights of Smith's collection using his own words as presented to his fellow collectors at the Hobby Club.

This chapter will draw on Smith's own words, taken from presentations and journal articles, and invite the reader to step inside the collection, as it existed in the early decades of the twentieth century.

4.2 Printed Material

The earliest printed items include the first printed edition of Euclid's *Elements* from 1482, Pacioli's *Summa de arithmetica, geometrica, proportioni et proportionalita* of 1494, and Boethius' *Opera* of 1499 (Frick, 1936b). Smith described his printed material as "mathematical classics, such as the first great printed algebra, the second one, the third, and the fourth. The Bombelli work belonged to the grandson of the author... I have also the first editions of all the Hindu classics in mathematics" (Smith, 1917, p. 12). Smith's collection of printed books and pamphlets later came to comprise about 10,000 pieces of printed material with texts ranging from the fifteenth through the sixteenth and seventeenth centuries.

In the course of his travels, Smith had also acquired one of the half-dozen copies with handwritten notes of Guillaume Libri's *History of Mathematics in Italy*, saved from a fire that consumed the remaining first editions in 1835 (Smith, 1936a). Smith also included in his collection major Japanese and Chinese mathematical texts, such as "the Chinese encyclopedia of mathematics pub-

4. Step Inside Smith's Collection

lished by the Jesuit influence in the seventeenth century; the first Chinese edition of Vlacq's table of logarithms; an early Chinese edition of Euclid; numerous Japanese manuscripts and printed works, and an early Manchu treatise on mathematical astronomy" (Educational Museum of Teachers College, 1909, para. 7).

During a trip to Florence in 1908, Smith purchased a copy of *Fabrica et uso del compasso di proportione* (1685) by Paolo Casati (1617–1707). This text demonstrates the beginnings of decimal fractions on its plate (Figure 4.2.1). During that same visit, Smith purchased a book on astrology that had a drawing of a compass rose signed by Galileo hidden inside (Lee, 2002). Another interesting find was the first German edition of John Napier's (1550–1617) *Rhabdologia Neperiana* of 1617. *Das ist newe und sehr leichte art durch etliche stäbichen allerhand zahlen ohne mühe, und hergegen gar gewiss zu multipliciren und zu dividiren* (1623), included the "Napier's Bones" version of multiplication (Lee, 2002).

Some of the printed materials included special autographs. For example, Tonstall's *De Arte Supputandi* (1529) with the autograph of Thomas Digges (1546–1595), an English mathematician (Figure 4.2.2). Smith also had "the proof sheets of Lord Brougham's address on Newton, with his corrections; the copy of Leslie's *Plane Trigonometry*... [which] he gave to Arago, the great French astronomer" (Smith, 1917, p. 13), and Johann Stoeffler's (1452–1531) *Elucidatio fabricae ususque astrolabii* (1513) that included diagrams depicting lines in an astrolabe plate with fold-out tabs. This work on the astrolabe was considered the standard up until the early seventeenth century. Another to point out is a first edition of Gemma Fresius' *Arithmeticae practicae methodus facilis* (1540) where the title page depicts Fresius in his study (Lee, 2002; Figure 4.2.3). Smith revealed his humorous side when he recreated this image, using himself rather than Fresius, for his own bookplate (Figure 4.2.4).

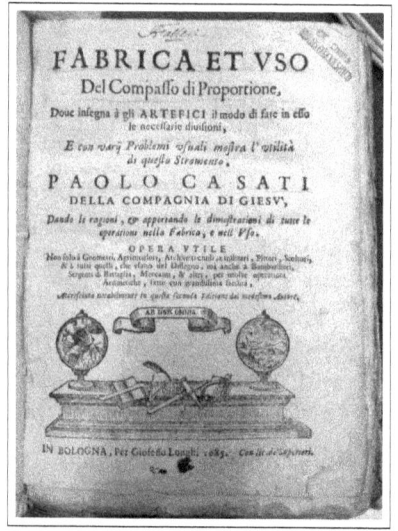

a. Title page.　　　　　b. Sample page including decimal fractions

Figure 4.2.1: Paolo Casati's *Fabrica et uso del compasso di proportione* (1685). Smith 510.78 1685 C26. Rare Book and Manuscript Library, Columbia University.

Figure 4.2.2: Tonstall's *De Arte Supputandi* (1529) with the autograph of Thomas Digges. On the left hand page, George A. Plimpton, who presented this text to Smith, signed it. Smith 510 1529 T83. Rare Book and Manuscript Library, Columbia University.

4.3 Manuscripts

Smith believed that the mathematical manuscripts from all over the world were among the most interesting items in his collection

4. Step Inside Smith's Collection

Figure 4.2.3: Gemma Fresius' *Arithmeticæ practicæ methodus facilis* (1540) title page. Smith 910 1540 Ap34. Rare Book and Manuscript Library, Columbia University.

Figure 4.2.4: Smith's bookplate (1907), modeled after Fresius' bookplate where Smith used his own image instead. Image courtesy of the Rare Book and Manuscript Library, Columbia University.

(Smith, 1917). He had a complete set of the Chinese classics, with special focus given to an early copy of a translation of Euclid by Matteo Ricci circa 1600. Ricci was a celebrated Jesuit missionary in China, whose name was deciphered by Smith as mentioned previously (Smith, 1936a). Smith considered the Japanese manuscripts in his collection beautiful both artistically and mathematically (Educational Museum of Teachers College, 1909).

In his talks and lectures, Smith highlighted the importance of mathematical manuscripts, in particular, ones from Persia, as "they translated the Greek classics into their own languages at a time when these classics were in danger of being utterly lost in the West. Thus it came about that the first knowledge that awakening Europe in the twelfth and thirteenth centuries had of writers, like Euclid, came through translations from the Arabic into Latin" (Smith, 1917, p. 14). Smith had collected many of these manuscripts, including, a "manuscript of Euclid's complete works of 1348 (747 of the Hegira), the two manuscripts of El Tusi's [Nasir al-Din al-Tusi] work, one of 1297 and the other of 1352, and several early manuscripts of Beha Eddin, the prince-mathematician of the sixteenth century" (Smith, 1917, p. 14–15).

Before Smith had translated the *Rubaiyat* (Omar, Smith, & Hussein, 1933) and had been honored by the Persian government in 1933, he described the manuscript and others in 1917 as "a Persian manuscript of the greatest Hindu classic, the *Lilavati* of Bhaskara [from the twelfth century], and while I have only the printed edition of Omar Khayyam's algebra (a manuscript belonging to me being still in India), you may be interested to see a beautiful little manuscript of his *Rubaiyat*" (Smith, 1917, p. 15). Another interesting manuscript was an unpublished life of Galileo Galilei (Educational Museum of Teachers College, 1909; Figure 4.3.1). Smith collected other types of manuscripts as well; these documents include "tax and census rolls...deeds...marriage con-

4. Step Inside Smith's Collection

Figure 4.3.1: Unpublished manuscript of Galileo's life. MS 520.1800. Description: "Manuscript written in the late 18th century. Apparently the author was a contemporary of Galileo. Numerous errors corrected in red, perhaps by Ferdinando Jacoli, to whom this ms. probably belonged." Rare Book and Manuscript Library, Columbia University.

tracts... units of measure for various cities, rent rolls and wills" (Frick, 1936b, p. 80).

4.4 Portraits and Medallions

Smith's collection of portraits contains approximately 3,000 portraits of mathematicians. In 1917, Smith wrote, "my collection of portraits of mathematicians started some thirty years ago in a desire to extra-illustrate Cantor's German *History of Mathematics*. It soon grew beyond this ambition, and now it numbers about 2,500 titles. These are rarely important as works of art, but they form an historical collection that is unique" (Smith, 1917, p. 11). Several of the portraits are black and white etchings; one of Guillaume de l'Hôpital (1661–1704) is unique in that it includes some color (Figure 4.4.1). The Department of Mathematics at Teachers College produced many lantern slides from these portraits. A complete list of the portraits currently at the Rare Book and Manuscript Library is included in Appendix D.

Smith also collected medals that honor mathematicians. Smith stated, "[i]t does not seem as if a mathematician would ever be looked upon with sufficient favor to have a medal struck in his

Figure 4.4.1: Portrait of Guillaume de l'Hôpital. Smith Portraits (Box 6). Rare Book and Manuscript Library, Columbia University.

honor, and yet I have about a hundred and fifty such evidences of the popularity of certain members of the guild" (Smith, 1917, p. 12). Among this collection are the images of Newton, Descartes, Fermat, Galileo, Neudorfer, Bertrand, Arago, and Le Verrier (Department of Mathematics of Teachers College, n.d.). The collection also included a complete set of mathematical portrait medallions by French sculptor David d'Angers (1788–1856) (Smith, 1917).

4.5 Autographs

Along with the autographed copies of texts, Smith collected letters from notable mathematicians. Yet again, his desire to "extra-illustrate" a book led him into the world of collecting letters (Smith, 1917). His collection consists of more than 4,000 items. In

4. Step Inside Smith's Collection 63

some instances, Smith collected all known correspondence of the author. According to Frick, all the most notable mathematicians and scientists are included in his collection (Frick, 1936b). These include Newton, Leibniz, Mersenne, and the families of Cassinis and Bernoullis. Smith described some quite interesting and rare finds of correspondence in his collection as:

> Letters written from the field by Delambre when he was surveying for the purpose of finding the basis of the metric system; a letter from Mechain, the unfortunate scientist who made the error in the survey which affected the length of the meter; letters from Libri relating to the serious charge of theft made against him in forming his two remarkable libraries; letters from Lewis Carroll of *Alice in Wonderland*, who was Charles Lutwidge Dodgson the Oxford tutor in mathematics (Figure 4.5.1); the letter written by Daniel Bernoulli to the Secretary of the French Academy in acknowledging the prize for his work on the tides; a page of Newton's manuscript; a love letter written by the great French mathematician Dupin; a poem written by Sylvester, England's great mathematician who gave the first real start to our science in America. (Smith, 1917, p. 16)

Other autographed letters in Smith's collection are from Sir William Rowan Hamilton, Euler, and an interesting chain of letters between Poncelet, Liouville, Direchlet, and Arago (Smith, 1917).

4. Step Inside Smith's Collection

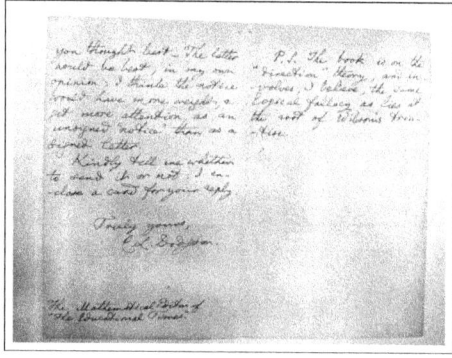

a. Page 1

b. Page 2

Figure 4.5.1: Letter from Charles Lutwidge Dodgson (Lewis Carroll) to the Mathematical Editor of "The Educational Times." David Eugene Smith Collection of Historical Papers [ca. 1400-1899]. (Box 9). Rare Book and Manuscript Library, Columbia University.

4.6 Instruments

Smith's collection of mathematical and astronomical instruments included about 275 pieces, which he often used in his lectures (Smith, 1936b). The texts and manuscripts Smith collected throughout the years included descriptions of mathematical tools and instruments; thus his text collection was directly related to his instruments. Some of the instruments in Smith's collection, especially those related to astronomical work, are quite beautiful. For example, his collection contains two seventeenth-century celestial spheres of bronze, where the stars are inlaid with silver (Appendix E: Smith 27-198, Smith 27-244). This type of instrument "represents a phase in the progress of the world from fear of the influence of the unknown to the appreciation, through mathematics, of what the stars really are, and where they are, and how they move, and what their substance is" (Smith, 1917, p. 4). A Japanese papier maché celestial sphere circa 1600 is also in Smith's collection (Appendix E: Smith 27-200).

Smith collected numerous armillary spheres—a mechanical model of the universe, quadrants, and astrolabes (Figure 4.6.1). One notable example of these is an Italian astrolabe from 1558 signed by Bernard Sabeus of Italy. Inscribed twice on the astrolabe is the signature of Sabeus; thus, he was quite proud of his work (Smith, 1917). Smith described these instruments as "used for measuring the angles of the stars above the horizon, for measuring angles from star to star, for determining the seasons and the latitude, for leveling, for running lines, and for the varied purposes for which we use our transit instruments today" (Smith, 1917, p. 5). Another astrolabe dear to Smith was a Hindu piece. In 1917 Smith described it as "only about 150 years old," but of "particular interest because it was bought by me from the last of a noted family of royal astrologers and was used as the model in the

 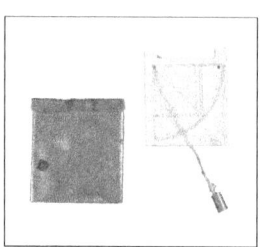 

a. Armillary sphere. Italy, circa 1550. Smith 27-197.

b. Quadrant. Italian, ivory with brass cover, early 19th century. Smith 27-246.

c. Astrolabe. Italian, signed by Bernard Sabeus, 1558. Smith 27-255.

Figure 4.6.1: Images courtesy of the Rare Book and Manuscript Library, Columbia University.

restoration of one of the largest astrolabes in a great Indian observatory [in Jaipur]" (Smith, 1917, p. 6). This subsection of Smith's collection contains three pairs of bronze compasses from Roman tombs (Figure 4.6.2). Smith thought these were quite interesting since one pair indicates, "one form of proportional compasses was as well known to ancient draftsmen as to those of our day" (Smith, 1917, p. 7).

Figure 4.6.2: Ancient Roman compass, about the beginning of the Christian era. Smith 27-286. Rare Book and Manuscript Library, Columbia University.

Although the instrument section of his collection consisted of only about 275 pieces, he was quite broad in his collecting. Smith was very interested in ancient dice (Figure 4.6.3). He collected

4. Step Inside Smith's Collection 67

seventy different dice, including, items "from the Etruscan tombs with the pre-Roman arrangement of dots, pieces from the period of the Persian invaders of Greece, pieces that show the transition from the primitive knuckle bones to the cubical form, glass dice from Egypt, icosahedral dice of the Ptolemaic period, and loaded pieces of the Roman gambling houses" (Smith, 1917, p. 7). He collected counters, some dating back to the Romans. As Smith explained, "the Roman boy carried his bag of calculi [counters] to school as some of us carried our slates" (Smith, 1917, p. 9). This method of counting continued as the universal method in Europe until the eighteenth century. Smith also collected pieces displaying the Roman numerals, such as tesserae, which are similar to counters but were used as "tickets" for admittance to games and performances (Smith, 1925a; Figure 4.6.4).

Smith's collection included numerous examples of ancient computing devices, such as Chinese bamboo rods, that were replaced around the thirteenth century by the suan-pan, which is the Chinese version of an abacus. The "rod" type of computing was translated to Japan in the form of the Japanese sangi (Figure 4.6.5), which was followed by the Japanese saroban, another type of abacus. Other examples in Smith's collection include Korean bones, Russian schoty, Napier rods, and the Armenian abacus (Smith, 1917).

Related to the ancient computing devices are methods for recording results, like the tally stick. Smith had some of these dating back to 1296 (Figure 4.6.6). This form of recording was used in many countries, but in England, counting sticks had an interesting history. Smith recounts:

> [In England], these tally sticks were used for centuries, and it [was] only within a hundred years that an act of Parliament [in 1834] ordered the vast accumulation of these wooden [sticks] burned [as there were no need

68   4. Step Inside Smith's Collection

a. "Roman. Gypsum, rudely made. Opposite sides marked 1–6, 3–5, 2–4, perhaps indicating that the piece is rather old. Not exactly Etruscan marking. Found near Civita Castellana, 43 miles from Rome." Smith 27-155.

b. "Roman. Bone, discolored by lying next to a piece of bronze. Correctly marked. Found at Francati." Smith 27-156.

c. "A hexagonal prism, bone, with an ivory handle for twirling. Nuremberg, c. 1800." Smith (250) 242.

d. Showing the three parts of the die to the left. Smith (250) 242.

Figure 4.6.3: Examples of dice. Descriptions from exhibition labels circa 1930. Rare Book and Manuscript Library, Columbia University.

4. Step Inside Smith's Collection

Figure 4.6.4: Examples of tesseræ with descriptive labels. David Eugene Smith Collection of Mathematical Instruments (Box D6). Rare Book and Manuscript Library, Columbia University.

Figure 4.6.5: "Japanese sangi sticks used in solving equations in the Old Japanese algebraic system. Purchased in Kyoto, Japan, 1907." Description from exhibition label circa 1930. Smith 27-292. Rare Book and Manuscript Library, Columbia University.

> for them due to more modern methods]. The carrying out of this law resulted in the burning of the Houses of Parliament, so [to this piece of calculation] we owe the beautiful piece of modern Gothic now overlooking the Thames. (Smith, 1917, p. 10)

As a result of the fire, English tally sticks are extremely rare.

Other items in Smith's collection include compasses, protractors, diagonal and gauger's scales, sundials (Figures 4.6.7 and 4.6.8), calendar medals, calendar rolls, eighteenth century drawing instruments, and a Ramsden telescope from 1775 (Figure 4.6.9). Items that depicted the magic square interested Smith as well (Figure 4.6.10). Smith also included some modern calculating machines, such as the Goldman and Stanley arithmometers, slide rules, and other adding machines (Department of Mathematics of Teachers College, n.d.).

A complete list of the instruments currently at the Rare Book and Manuscript Library is included in Appendix E.

4. Step Inside Smith's Collection

a. Leather covered, silk lined decorative box.

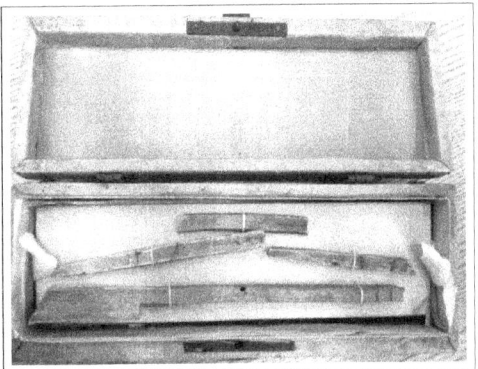

b. Four wooden tally sticks.

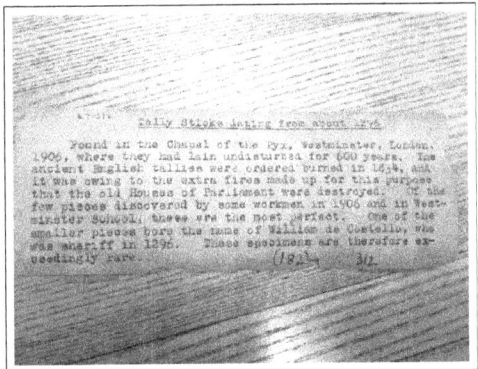

c. Descriptive label used when exhibited.

Figure 4.6.6: English tally sticks of 1296. Smith 27-312. Rare Book and Manuscript Library, Columbia University.

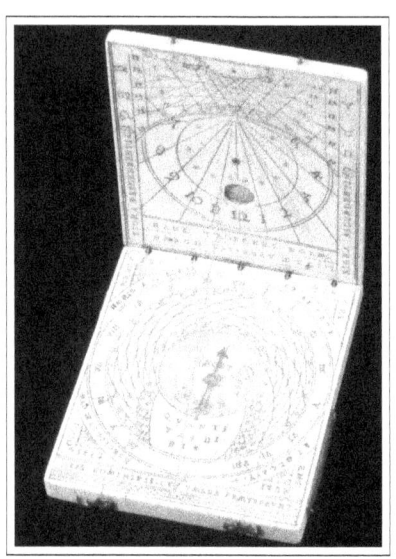

Figure 4.6.7: "Nuremberg, signed by Hans Tr chel, 1603. Ivory, with string gnomon horizontal dial and pin gnomon for vertical dial." Description from exhibition label circa 1930. Smith 27-225. Image courtesy of the Rare Book and Manuscript Library, Columbia University.

Figure 4.6.8: "Cubical sundial. Bavarian, 18th century. Horizontal and vertical. North, south, east, and west." Description from exhibition label circa 1930. Smith 27-222. Image courtesy of the Rare Book and Manuscript Library, Columbia University.

4. Step Inside Smith's Collection

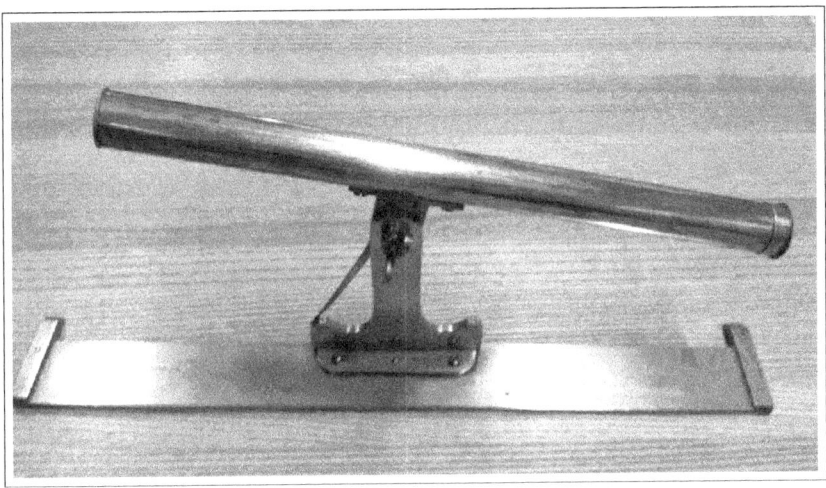

Figure 4.6.9: Ramsden Telescope. "Telescope said to have been made by Ramsden of London, the great maker of mathematical instruments about 1775." Description from exhibition label circa 1930. Smith 27-267. Rare Book and Manuscript Library, Columbia University.

"Magic square on reverse of medal showing Venus (contains errors)." Smith 254a.

"Arabic(?) Amulet, found at Karnak. Illustrates the degeneration of the Magic Square." Smith (253). 27-311

"Amulet. Christian-Kabbalistic. On the rim are Greek and Hebrew words separated by a cross. Magic square that adds to 175." Smith 27-318.

Figure 4.6.10: Representations of the Magic Square. Descriptions from exhibition labels circa 1930. Rare Book and Manuscript Library, Columbia University.

4.7 Two Exhibition Pamphlets

Two pamphlets, one from the Department of Mathematics and another from the Educational Museum of Teachers College, provide another rare glimpse of the contemporary view of Smith's collection. These pamphlets are currently preserved in the David Eugene Smith Professional Papers collection in the Rare Book and Manuscript Library at Columbia University. Their text are reproduced here with permission. The pamphlet from the Department of Mathematics is a general description of all that is included in Smith's collection. This pamphlet would have been the likely starting point for visitors who came to Teachers College to view the collection. It may also have been circulated to other colleges as a way to promote the collection. Unfortunately, it is not dated, but it can be assumed that the pamphlet was produced around 1909, as it is very similar in design to the Educational Museum pamphlet of that year.

As a guide for visitors to the Educational Museum of Teachers College, the second pamphlet mentions an exhibition from January 4, 1909 through February 13, 1909 of Smith's collection during that time. A special activity of this exhibit was Smith's presentation and explanation of his own material on January 9th. This exhibit consisted of:

> Mathematical instruments, measures, medals, manuscripts, early printed books, portraits, and curios, collected in various parts of the world and illustrating the history and teaching of mathematics in various periods. There are also photographs of many rare manuscripts and early printed works in various libraries of Europe and America, supplementing the original material in the collection. (Educational Museum of Teachers College, 1909)

The pamphlet is organized by the cases in the actual Museum. It describes the contents of twenty-six cases, as well as noting that there are additional materials scattered around the walls and that the full collection includes many additional objects of interest:

> Owing to the lack of room in the museum it is impossible to exhibit a great many books and objects that should supplement what is here displayed. These include books showing the early history of the calculus and analytics, portraits, autographs, photographs of rare inscriptions, and illustrations of primitive instruments. (Educational Museum of Teachers College, 1909)

4.8 References

Buffalo Society of Natural Sciences. (1910). *Bulletin of the Buffalo Society of Natural Sciences*, $X(1)$.

Columbia University Quarterly. (1909). Teachers College. Columbia University: New York.

Columbia University Quarterly. (1910). Teachers College. Columbia University: New York.

Department of Mathematics of Teachers College. (n.d.) [Pamphlet describing David Eugene Smith's Collection]. David Eugene Smith Professional Papers, 1860-1945 (Box 137). Rare Book and Manuscript Library, Columbia University, New York.

Educational Museum of Teachers College. (1909) [Pamphlet describing the Exhibition of Material Illustrating the Historical Development of Mathematics from the collection of David Eugene Smith]. David Eugene Smith Professional Papers, 1860-1945 (Box 137). Rare Book and Manuscript Library, Columbia University, New York.

Frick, B. M. (1936b). The David Eugene Smith Mathematical Library of Columbia University. *Osiris* $I(1)$, 78-84.

Hobby Club. (1920). *Annals of the Hobby Club of New York City, 1912-1920.* New York: The De Vinne Press.

Lee, J. (2002). [Master list of items in 2002 exhibit of David Eugene Smith's collection]. Rare Book and Manuscript Library, Columbia University, New York.

New York Museum of Science and Industry. (1930). *Exhibit of early astronomical and mathematical instruments.* New York: Museums of the Peaceful Arts.

Omar, K., Smith, D. E., & Hussein, H. (1933). *The Rubáiyát of Omar Khayyám.* New York: B. Westermann Co.

Peters, G. M. (1911, May 16). [Letter to David Eugene Smith]. David Eugene Smith Professional Papers, 1860-1945 (Box 38). Rare Book and Manuscript Library, Columbia University, New York.

Smith, D. E. (1911, April 12). [Letter to C. T. McFarlane]. David Eugene Smith Professional Papers, 1860-1945 (Box 32). Rare Book and Manuscript Library, Columbia University, New York.

Smith, D. E. (1917). [Notes on talk, Mirabilia Mathematica, given to Hobby Club]. David Eugene Smith Professional Papers, 1860-1945 (Box 88). Rare Book and Manuscript Library, Columbia University, New York.

Smith, D. E. (1925a). *History of mathematics: Volume II.* Boston: Ginn and Company.

Smith, D. E. (1936a). [Unpublished little folding book]. David Eugene Smith Professional Papers, 1860-1945 (Box 115). Rare Book and Manuscript Library, Columbia University, New York.

Smith, D. E. (1936b). The David Eugene Smith gift of historical-mathematical instruments to Columbia University. *Science 83* (2143), 79-80.

Wade, H. T. (1909). A museum to illustrate the development of mathematics. *Scientific American 101*(10), 10-19.

Chapter 5

Smith's Collection on the Move

Just ten years after its founding in 1899, and while the Educational Museum of Teachers College was proudly showcasing Smith's collection in the exhibition of 1909, the museum was approaching a crisis point. The size of the collection, in part due to Smith's own zeal as an avid collector of mathematics manuscripts and artifacts, was rapidly expanding beyond the space that Teachers College could make available to house it.

As early as 1907, Dean Russell alerted Smith to the inevitability of a space shortage. A letter to Smith from Russell on October 11, 1907 notes that, "I am hoping to provide in the museum more rooms for your jimcrackery. However, while this new building will not give the relief counted on, in some respects it will make things much more comfortable" (Russell, 1907, p. 1). In a sign of his continuing support for the museum's mission, Dean Russell persistently attempted to create more space for Smith's collection. Russell's 1909 *Report of the Dean* included a section on the Educational Museum, noting that "with more floor space next year we shall make available for student use the extensive private collections of Professor Smith on the history of mathematics and of Professor Monroe on the history of education" (p. 12). It is clear from the 1909 exhibit catalog notes that Smith's collection was already too large to fit into the combined space of the Mathematics Department and the main room of the Educational Museum. While Russell's support for the Museum and its growing collection seems to have been quite genuine, his reference to Smith's prized artifacts as "jimcrackery" is telling, particularly in the context of the historic reluctance of Columbia's Trustees to welcome the integration of Teachers College. It is not surprising, therefore, to learn that even while Russell and Smith petitioned for more administrative support and collection space, the growth period of the Educational Museum was already drawing to a close.

The academic expansion of Teachers College in the years after 1909 meant that more and more lecture rooms were needed to hold the growing student population. Providing classroom space rather than exhibition rooms for the Museum was a clear administrative priority. As the possibility of any significant Museum expansion faded, Smith tried to convince the administration that this lack of consideration of adding more exhibit space was a huge disservice to the students and the integrity of the mission of Teachers College. With strict space constraints, the museum was not able to purchase new material, as any new acquisitions would have nowhere to be displayed. Thus, the collection was becoming dormant. Smith believed that after its promising first decade, Teachers College was falling behind just as other institutions were moving forward with their museums and collections (Smith, 1913).

Smith finally acknowledged that the much-needed museum expansion space was not going to be available in the foreseeable future. In a 1913 letter to Russell, Smith took the initiative to recommend the closing of the Educational Museum. "It is with much regret that I am compelled to recommend the closing of the Educational Museum until the time shall come when we have room for an adequate display of our material" (Smith, 1913, p. 1).

5. Smith's Collection on the Move

Smith remained the supervisor of the collection until 1914 when the museum dispersed its collection and closed.

The 1913-14 *Teachers College School of Education Announcement* publicized that "as this announcement is going to press the College has just made arrangements to transfer its museum interests to the Permanent Educational Exhibit recently established by Mr. George A. Plimpton at Fifth Avenue and Thirteenth Street where there is installed in special rooms the greatest exhibit of modern school equipment ever brought together besides a valuable exhibit of historical material relating to education" (p. 123). Thanks to their long-standing friendship, Plimpton agreed to make space available in his new educational venture for Smith's private mathematics collections (Neilson, 1913).

Even at this point, however, Smith continued to hope that Columbia and Teachers College would not completely abandon the mission embodied in the Educational Museum and his vision of a dynamic academic collection of historical artifacts that could be used in the teaching of mathematics. Although the Educational Museum of Teachers College was officially closed, Smith urged that it should be resurrected at some future date pointing to the recognition which the Museum had achieved in its short life. In a letter on November 3, 1915, Smith urged Dean Russell to give some consideration to reinstating a bigger and even more ambitious museum at Teachers College:

> I wish to also call attention to the need for a first-class educational museum. The one which we had was becoming well known, and with a good curator it would have been of great value to the College and to the profession in general. It was visited by distinguished educators and was frequently mentioned in educational articles. Such a museum requires, however, at least twice the space which ours had, and it requires the entire time of an in-

telligent curator. We have nothing in this country which meets the need, nothing, for example, as good as the educational collections in Dresden, Paris, and Leipzig, to mention those with which I am fairly familiar. The need for a fire-proof building for the library, museum, and executive offices is very pressing. (Smith, 1915, p. 1)

Smith kept up his advocacy for reviving the Museum in a letter to Russell on February 19, 1916. Smith stated, "I have attended to the packing up of other museum material and storing it in the basement awaiting the time when we may have space and money for a large educational museum such as this college certainly ought to have" (Smith, 1916a, p. 1).

In the meantime, the part of the collection which had found a home at Plimpton's new venture was not destined to stay in this location for long. Despite having abundant space (Plimpton owned entire floors in the building where the Permanent Educational Exhibit Company was located) and the support of Sarah Mitchell Neilson, Smith's former assistant at the Educational Museum, who became the secretary of the Permanent Educational Exhibit Company, this company was not a commercial success. Notwithstanding its optimistic name, the Permanent Educational Exhibit, which focused on displaying modern educational material for use by teachers and possible sales, closed towards the end of 1917. This was devastating for Plimpton, Neilson, and Smith, as they had hoped that it would be a great success (Neilson, 1917).

With the closure of the Educational Museum of Teachers College in 1914, the dissolution of Plimpton's Permanent Educational Exhibit in 1917, and no sign of interest from Teachers College to invest in a new, expanded museum space, Smith shifted his focus to the display and housing of his own private collection. These materials had remained in Smith's possession in the Department of Mathematics as he used items from the collection extensively in

5. Smith's Collection on the Move

his lectures and research. During the next decade, Smith worked to make his personal collection as accessible as possible. He loaned rare items to feature in special exhibits at the Bryson Library and Columbia's Libraries (Upton, 1916; Refior, 1924; Williamson, 1926). He also established a relationship with yet another emerging New York institution that showcased historical artifacts—the Museums of the Peaceful Arts.

George F. Kunz (1856–1932) was the president of the Association for the Establishment and Maintenance for the People in the City of New York of Museums of Peaceful Arts. In 1912, at the annual meeting of the American Association of Museums, he proposed the formation of twenty museums dedicated to the industrial arts, which would be named the Museums of the Peaceful Arts. This plan was enacted in 1913 and continued until the 1930s, with the headquarters at 24 West 40th Street (Kunz, 1927). Smith was acquainted with Kunz through their mutual membership in the Hobby Club, which he had joined in 1913 (Hobby Club, 1920). Members included Kunz, a committee member and gem enthusiast, Plimpton with his passion for early educational books and manuscripts, and many other notable gentlemen (The New York Times, 1920).

In 1926 Smith became actively involved in the Museums of Peaceful Arts, arranging for an exhibit of calculating machines. In organizing this exhibit, he asked Clifford B. Upton, then the secretary of Teachers College and eventually a leader in the Department of Mathematics of Teachers College, what would be an appropriate type and size of exhibition for his materials (Upton, 1926). Smith continued to exhibit portions of his collection at the Museums of Peaceful Arts exhibit, and in 1928 he convinced Ernest G. Yalden, an expert in sun dialing, to write a monograph on the subject that would include Smith's collection.

Smith wrote to Yalden in 1928 regarding this matter, "I shall be glad to meet you at the Museums of the Peaceful Arts on Tuesday, January 31st, at one-thirty. The temporary rooms are at 24 West 40th Street, and I shall be on the seventh floor. I suggest that special place because my material is displayed there, and we can talk over both propositions better than at any other place" (Smith, 1928, p. 1). In a letter to Smith from Kunz dated February 20, 1928, Smith's collection was again recognized as being one of a kind: "We believe this is the only exhibit of this kind that has been made in this country up to date and are very happy to have it in our museum" (Kunz, 1928, p. 1). Yalden's paper was finally published in 1930, along with an essay by Jekuthiel Ginsburg on astrolabes and a catalogue of Smith's mathematical instruments.

Upton believed that the catalog of instruments in Smith's collection would be "an important reference book for all of us" (Upton, 1930, p. 1). Unfortunately, the Museums of Peaceful Arts was another institution that did not establish long-term roots in New York; it disbanded in the 1930s. The collections exhibited in the Museums of Peaceful Arts, including Smith's exhibit of mathematical instruments, were transferred to the New York Museum of Science and Industry (Shaw, 2011). No doubt this period of short-lived institutions and closed exhibits that included a decades-long migration of his collection was frustrating to Smith. In a letter to "an inquiring mind" regarding the fate of the Educational Museum of Teachers College and his collection of mathematical instruments dated July 26, 1934, Smith simply stated, "I gave all that material to the [New York Museum of Science and Industry] some years ago" (Smith, 1934a, p. 1).

Smith did not, however, give away the preponderance of his historical collection of mathematics texts and artifacts during the 1920s and 1930s, any more than he ever gave up on his vision of sharing these materials with mathematics teachers and students worldwide. Smith's decisions about how to achieve this goal after

5. Smith's Collection on the Move   85

the closure of the Educational Museum of Teachers College is the topic of the next chapter.

5.1 References

Hobby Club. (1920). *Annals of the Hobby Club of New York City, 1912-1920*. New York: The De Vinne Press.

Kunz, G. F. (1927, April 4). [Letter to Jekuthiel Ginsburg]. David Eugene Smith Professional Papers, 1860-1945 (Box 30). Rare Book and Manuscript Library, Columbia University, New York.

Kunz, G. F. (1928, February 20). [Letter to David Eugene Smith]. David Eugene Smith Professional Papers, 1860-1945 (Box 30). Rare Book and Manuscript Library, Columbia University, New York.

Neilson, S. M. (1913, December 11). [Letter to David Eugene Smith]. David Eugene Smith Professional Papers, 1860-1945 (Box 36). Rare Book and Manuscript Library, Columbia University, New York.

Neilson, S. M. (1917, August 22). [Letter to David Eugene Smith]. David Eugene Smith Professional Papers, 1860-1945 (Box 36). Rare Book and Manuscript Library, Columbia University, New York.

New York Times. (1920, August 1). A club which justifies the man with a hobby. New York Times.

Refior, S. R. (1924). From the shelves of Dr. David Eugene Smith's unique mathematical historical library. *Mathematics Teacher, 17*(5), 269-273.

Russell, J. E. (1907, October 11). [Letter to David Eugene Smith]. David Eugene Smith Professional Papers, 1860-1945 (Box 41). Rare Book and Manuscript Library, Columbia University, New York.

Russell, J. E. (1909). Report of the Dean. *Teachers College Bulletin*. New York: Teachers College.

Shaw, D. (2011). The Museums of the Peaceful Arts: A timeless dream detailed in a curious set of scrapbooks. *Smithsonian Collections Blog*, Smithsonian Institution, Washington, D.C.

Smith, D. E. (1913, September 23). [Letter to James E. Russell]. David Eugene Smith Professional Papers, 1860-1945 (Box 41). Rare Book and Manuscript Library, Columbia University, New York.

Smith, D. E. (1915, November 3). [Letter to James E. Russell]. David Eugene Smith Professional Papers, 1860-1945 (Box 41). Rare Book and Manuscript Library, Columbia University, New York.

Smith, D. E. (1916a, February 19). [Letter to James E. Russell]. David Eugene Smith Professional Papers, 1860-1945 (Box 41). Rare Book and Manuscript Library, Columbia University, New York.

Smith, D. E. (1928, January 28). [Letter to Ernest G. Yalden]. David Eugene Smith Professional Papers, 1860-1945 (Box 47). Rare Book and Manuscript Library, Columbia University, New York.

Smith, D. E. (1934a, July 26). [Letter to K. P. Morgan]. David Eugene Smith Professional Papers, 1860-1945 (Box 35). Rare Book and Manuscript Library, Columbia University, New York.

Teachers College School of Education Announcement. (1913). Teachers College, New York.

Upton, C. B. (1916, February 18). [Letter to David Eugene Smith]. David Eugene Smith Professional Papers, 1860-1945 (Box 52). Rare Book and Manuscript Library, Columbia University, New York.

Upton, C. B. (1926, December 2). [Letter to David Eugene Smith]. David Eugene Smith Professional Papers, 1860-1945 (Box

52). Rare Book and Manuscript Library, Columbia University, New York.

Upton, C. B. (1930, July 30). [Letter to David Eugene Smith]. David Eugene Smith Professional Papers, 1860-1945 (Box 52). Rare Book and Manuscript Library, Columbia University, New York.

Williamson, C. C. (1926, February 21). [Letter to David Eugene Smith]. David Eugene Smith Professional Papers, 1860-1945 (Box 54). Rare Book and Manuscript Library, Columbia University, New York.

Chapter 6

A Collection Without Walls

The multiple homes and exhibits of Smith's collection, including the Educational Museum of Teachers College and the Department of Mathematics of Teachers College, the Permanent Educational Exhibit, the Museums of Peaceful Arts, and the New York Museum of Science and Industry, were only one facet of the lasting impact of Smith's collection on mathematics education. From his early career, Smith actively used the items in the collection to enhance his own teaching and research, to illustrate his publications, and to support other mathematics education programs.

In Smith's 1936 description of his donation of mathematical instruments to Columbia University, he stated that he used his collection, especially the instruments, in his lectures (Smith, 1936b). Smith felt that a proper program for mathematics education ought to have a history of mathematics course. Moreover, such a course should include opportunities for students to work with primary sources and artifacts. Smith's desire to foster the design of such courses and his generous nature motivated him to provide access to his collection through beautifully illustrated texts, published lectures, and traveling exhibits, as well as through an important technology of his time—sets of lantern slides.

6.1 The Lantern Slides

Stereopticon slides, also called sciopticon, magic lantern, or lantern slides, were popular for about one hundred years—beginning in the mid-nineteenth century. By 1880, the stereopticon slides were becoming more popular in colleges and universities because a machine had been developed to mass-produce the slides (Spindler, 1988). Not until 1907 did Smith and the Educational Museum produce a set of slides based on his and George A. Plimpton's collections. The series of slides was called "Illustrations for Lectures on the History of Mathematics," reproduced in Appendix A, and were made by James Huntington, 610 St. Johns Place, Brooklyn, NY (Smith, 1907).

Through lantern slides, Smith's collection had a direct impact on mathematics students throughout the United States. Upon request, Smith produced sets of lantern slides to be sent out to institutions and individuals. Due to the size of Smith's collection, the Museum's staff felt confident creating sets in "nearly every branch of the subject" (Smith, 1907, p. 1). Some of these slides still exist in schools and libraries around the nation, including more than 150 slides kept at the current Mathematics Program of Teachers College.

Smith, however, must have produced his own set of slides for his teaching before that since, in a letter to Smith dated June 5, 1905, Louis Charles Karpinski (1878–1956) stated, "I thank you very heartily for consenting to loan me a set of the 'mathematical slides.' I assume the responsibility for their safe return. I realize that the set of slides is something quite unique not only in the teaching of the history of mathematics but in the teaching of any history" (Karpinski, 1905, p. 1). Creating slides to display historical information would soon become common practice during the early twentieth century. In 1933, Karpinski, an American mathematician and contemporary of Smith, later created his own set of slides for the Chicago Exposition where he had four series consisting of arithmetic, algebra, geometry, and trigonometry (Karpinski, 1933).

The correspondence in Smith's Professional Papers at the Rare Book and Manuscript Library includes numerous letters from professors and teachers asking for the slides. For example, Professor Arthur Gale of the University of Rochester wrote Smith on March 2, 1907, saying that he planned on using the lantern slides in his teaching of freshman; since it might interest those who are not planning on continuing in mathematics (Gale, 1907). On March 28, 1911, R. B. McClenon of the Mathematics and Astronomy Department of Grinnell College in Iowa wrote to acquire a set of slides; McClenon had read that the set was used in Smith's History of Mathematics course and he planned on teaching a similar course at Grinnell (McClenon, 1911). McClenon would later become the Librarian for the Mathematical Association of America, of which Smith was president in 1920. A complete collection of the first series of 119 slides commissioned by Smith is currently available at the University of Kansas (University of Kansas, n.d.).

After the Educational Museum closed in 1914, the slides could no longer be purchased. They were, however, able to be loaned inside of Teachers College and Columbia University. As Neilson, former assistant to the Educational Museum of Teachers College, wrote to Smith's secretary, Mrs. E. A. Mitchell, in 1914, "the rule was, and I suppose still is, that no slides can be loaned outside of the University. You will get into a peck of trouble if you do not keep to this, as I know to my sorrow" (Neilson, 1914, p. 1). Eventually a set of slides were kept exclusively in the Department of Mathematics rather than in Bryson Library. A letter dated December 11, 1933 from Clifford B. Upton and William D. Reeve to Smith notes that the stereopticon viewer would also be kept in the Department (Upton & Reeve, 1933).

The catalog of slides, "Illustrations for Lectures on the History of Mathematics" (1907), contains about 275 items. These are from both Smith and Plimpton's collections. The slides that are still held by the current Mathematics Program of Teachers College are published in Part Two of this book.

6.2 Texts

Another popular request in the correspondence of Smith's Professional Papers at the Rare Book and Manuscript Library of Columbia University is that of a series of portraits. Varying types of institutions sent in this request; for example, many universities, such as the University of Chicago, numerous secondary school teachers wished to acquire copies of these valuable portraits, and even a savings bank in Massachusetts (Montgomery, 1912).

This popularity was due to the fact that Smith had published with Ginn & Company in 1916 a series of portraits of notable mathematicians throughout history entitled *Mathematical Portraits and Pages* (Smith, 1916b). A few years earlier, Smith had published the *High School Portfolio of Mathematicians* with Open Court Publishing Company. It seems that these images were rather small. A letter from a schoolteacher in Worcester, Massachusetts, complained, "I purchased... 'High School Portfolio of Mathematicians' expecting to frame and hang on school room walls. I find the pictures so small that they are of little use for such purpose. I write you as editor of the portfolios to know if it is possible to get these pictures in larger size" (Parker, 1913, p. 1). As stated in Chapter 4, throughout Smith's travels he had obtained over 3,000 of such portraits.

As might be expected, Smith's collection of mathematical history was a source of inspiration for his own publications. He

6. A Collection Without Walls

wrote numerous entries for Professor Paul Monroe's *Cyclopedia of Education* (1911–1913), and also journal articles and pamphlets related to the history of mathematics that included direct links to his collection. A specific example of this was his three-part series published in the *American Mathematical Monthly* in 1925. In each installment of the series entitled "The Surnamed Chosen Chest," Smith listed major sub-sets in his own collection based on association copies, oriental works, and portrait medals (Smith, 1925b).

Smith's collection also served as a primary resource for *A History of Japanese Mathematics*, *Number Stories of Long Ago*, *The History of Mathematics Volumes I* and *II*, and *The Wonderful Wonders of One-Two-Three*. Within all of these texts, Smith beautifully illustrated the history of mathematics. All of the photographs in *A History of Japanese Mathematics*, co-authored with Yoshio Mikami, were taken from works in Smith's collection. As a final note in the Preface of said text, Smith wrote:

> It is only just to mention at this time the generous assistance rendered by Mr. Leslie Leland Locke, one of my graduate students in the history of mathematics, who made in my library the photographs for all of the illustrations used in this work. His intelligent and painstaking efforts to carry out the wishes of the authors have resulted in a series of illustrations that not merely elucidate the text, but give a visual idea of the genius of the Japanese mathematics that words alone cannot give. (Smith & Mikami, 1914, p. v)

Leslie Leland Locke eventually became a major collector of calculating machines. His collection was donated to the Smithsonian in 1939 (Leland Locke, 1939).

Smith's recognition of how a photograph, or any visual aid, can enormously affect the understanding of the history of mathematics is further displayed in *Number Stories of Long Ago* (1919) and *The Wonderful Wonders of One-Two-Three*. Both of these texts were meant for a younger audience, so perhaps the need for visualizations was even greater. Throughout these texts are numerous images and illustrations taken from Smith's collection as well as the collections of others, such as Plimpton. Smith understood that creating these books in this way would draw the reader in and make them interested in the subject.

The two-volume collection of *The History of Mathematics* was meant for students and teachers as a practical text for learning the historical side of mathematics (Smith, 1923). Smith commented to the reader regarding the illustrations throughout the book:

> In the selection of illustrations the general plan has been to include only such as will be helpful to the reader or likely to stimulate his interest. It would be undesirable to attempt to give, even it this were possible, illustrations from all the important sources, for this would tend to weary the reader. On the other hand, where the student has no access to a classic that is being described or even to a work, which is mentioned as having contributed to the world's progress in some humbler manner, a page in facsimile is often of value. (Smith, 1923, p. viii)

As with the distribution of the lantern slides, Smith was always considerate of the fact that not all students or teachers would have physical access to such a remarkable collection. Thus, providing images from his collection in his texts, articles, and other publications was a gift from Smith to the mathematics community.

6.3 Teaching and Mentoring

Throughout Smith's tenure at Teachers College he annually gave courses on the history of mathematics. These courses were very popular, and the students thoroughly enjoyed learning the history of mathematics through Smith's collection. One student's appreciation of his course is evident in an article written in the *Mathematics Teacher* in 1924. The student, Sophia Refior, stated how "the inspiration derived from examining these priceless objects" motivated her to publish the descriptions of some rare items in Smith's collection (Refior, 1924, p. 269).

Smith was dedicated to teaching a proper history of mathematics course. In fact, towards the end of his teaching career, he was teaching three sequences of year-long history of mathematics courses. In the first course, he used the usual lecture style to instill an understanding of the development of mathematics. In the second semester of the first level, students who showed strength in research were allowed access to Smith's collection for study of source material. The second course had the first course as a prerequisite. "The class was small, ordinarily made up of about a dozen students who had the scientific and linguistic ability to begin a year of pre-seminar work" (Smith, n.d.b, p. 4). This class was given direct access to Smith's collection, and since his collection contained a considerable amount of scholarly work in varying languages, these students had reading knowledge of at least two languages besides English. "In this course each student was asked to select a topic and to work upon it as long as he chose. Some worked upon three or four in a year; others continued the work upon a single topic through this and the seminar year" (Smith, n.d.b, p. 5). The third course was for individual work towards a doctoral dissertation, and the student worked with Smith individually to develop a thesis. After Smith officially retired in 1926, he continued to work with students—some of whom were not even registered at Teachers College—to produce publications based on Smith's collection (Smith, n.d.b).

Smith believed that other colleges and universities ought to provide similar courses, though he felt that they would be successful only if the library at those institutions had a collection of original material. He claimed:

> It is possible to secure this [original material] at a reasonable expenditure so far as early printed books are concerned, if one avoids incunabula and specially rare books. Any university library can afford to purchase fifty volumes of the 16th and 17th centuries, a set of the *Bibliotheca Mathematica*, one of Boncompagni's *Bulletino*, and the leading histories. Even the smaller college libraries can form a working collection at not too great expense. (Smith, n.d.b, p. 6)

Smith understood that libraries such as he and Plimpton had cultivated could not be duplicated; however, he suggested that portraits be acquired through travel abroad and commented that it took him forty years to collect the 3,000 portraits in his collection. Though the ancient mathematical instruments and rare manuscripts may have become too expensive for institutions to purchase, students could rely on the illustrations in Smith's texts for their research (Smith, n.d.b).

6.4 References

Gale, A. (1907, March 2). [Letter to David Eugene Smith]. David Eugene Smith Professional Papers, 1860-1945 (Box 18). Rare Book and Manuscript Library, Columbia University, New York.

Karpinski, L. C. (1905, June 5). [Letter to David Eugene Smith]. David Eugene Smith Professional Papers, 1860-1945 (Box 28).
Rare Book and Manuscript Library, Columbia University, New York.

Karpinski, L. C. (1933, May 4). [Letter to David Eugene Smith]. David Eugene Smith Professional Papers, 1860-1945 (Box 28).
Rare Book and Manuscript Library, Columbia University, New York.

Leland Locke, L. (1939, December 31). [Letter to David Eugene Smith]. David Eugene Smith Professional Papers, 1860-1945 (Box 31). Rare Book and Manuscript Library, Columbia University, New York.

McClenon, R. B. (1911, March 28). [Letter to David Eugene Smith]. David Eugene Smith Professional Papers, 1860-1945 (Box 28). Rare Book and Manuscript Library, Columbia University, New York.

Montgomery, W. J. (1912, August 9). [Letter to David Eugene Smith]. David Eugene Smith Professional Papers, 1860-1945 (Box 35). Rare Book and Manuscript Library, Columbia University, New York.

Neilson, S. M. (1914, November 17). [Letter to E. A. Mitchell]. David Eugene Smith Professional Papers, 1860-1945 (Box 36). Rare Book and Manuscript Library, Columbia University, New York.

Parker, W. P. (1913, December 4). [Letter to David Eugene

Smith]. David Eugene Smith Professional Papers, 1860-1945 (Box 38). Rare Book and Manuscript Library, Columbia University, New York.

Refior, S. R. (1924). From the shelves of Dr. David Eugene Smith's unique mathematical historical library. *Mathematics Teacher, 17*(5), 269-273.

Smith, D. E. (n.d.b) [Notes about methods for teaching the history of mathematics]. David Eugene Smith Professional Papers, 1860-1945 (Box 88). Rare Book and Manuscript Library, Columbia University, New York.

Smith, D. E. (1907). Illustrations for lectures on the history of mathematics. *Educational museum: Teachers College Columbia University, N.Y.*

Smith, D. E., & Mikami, Y. (1914). *A History of Japanese mathematics.* Leipzig: F. Meiner.

Smith, D. E. (1916b). *Mathematical portraits and pages.* New York: Ginn & Company.

Smith, D. E. (1919). *Number stories of long ago.* Boston: Ginn & Company.

Smith, D. E. (1923). *History of mathematics: Volume I.* Boston: Ginn & Company.

Smith, D. E. (1925a). *History of mathematics: Volume II.* Boston: Ginn and Company.

Smith, D. E. (1925b). The surnamed chosen chest. *American Mathematical Monthly, 32,* 287-294, 393-397, 444-450.

Smith, D. E. (1936b). The David Eugene Smith gift of historical-mathematical instruments to Columbia University. *Science 83*(2143), 79-80.

Smith, D. E. (1937). *The wonderful wonders of one-two-three.* New York: McFarlane, Warde, McFarlane.

Spindler, R. P. (1988). "Windows to the American Past: Lantern Slides as Historical Evidence." *Visual Resources, 5,* 1-15.

University of Kansas. (n.d.). Guide to the collection of educational lantern slides: Collection of lantern slides for the history of education and mathematics, 1900-[193?]. Kenneth Spencer Research Library, University of Kansas, Kansas.

Upton, C. B. & Reeve, W. D. (1933, December 11). [Letter to David Eugene Smith]. David Eugene Smith Professional Papers, 1860-1945 (Box 52). Rare Book and Manuscript Library, Columbia University, New York.

Chapter 7

Columbia University Receives a Gift

Although David Eugene Smith had traveled the world, his home and heart were at Teachers College and Columbia University. As Smith was growing older and his collection still did not have a permanent exhibition space, he collaborated with the libraries at Columbia University to make the historical collection available to students and interested professionals. Once again, as in the case of the Educational Museum of Teachers College, Smith would step up to initiate an association to promote the conservation of pieces with historical value.

7.1 Smith's Donation

Library Friends Associations began during the 1920s across the country and grew in popularity with each decade. In November 1928, Smith formed the "Friends of the Library of Columbia University." Smith was secretary, and he persuaded his friend and notable bibliophile George A. Plimpton to serve as president and Frank Fackenthal as treasurer. Fackenthal had a life-long association with Columbia University, as student, secretary, provost, and acting president from 1945-1948 (Hyde, 1971).

The purpose of the association (basically the purpose of all Friends groups) was to give supplementary aid beyond the yearly library budget, which, everywhere, provides barely enough money for the purchase of current books and journals. This is why there is always emphasis upon rare books and manuscripts—the material that scholars need and the library cannot afford (Hyde, 1971, p. 7).

As part of his association with the Friends group, Smith spearheaded the creation in 1933 of *Bibliotheca Columbiana*. He also wrote and edited the only four issues published, which promoted the activities of the libraries of Columbia and Teachers College. The Friends worked to solicit library support and donations from outside parties. Knowing that a considerable initial contribution would spark the interest of and perhaps inspire other donors, in 1931, Smith started with the donation of his own library of "mathematical works, Orientalia, medieval and renaissance documents and manuscripts, and letters and portraits of prominent mathematicians. The collection totaled some 20,000 pieces" (McAleer, 1961, p. 19). Some pieces were apparently already at Columbia since in 1930 Smith wrote Prince Damrong to assure him that the three manuscripts that were given to him for Columbia University had been delivered safely (Smith, 1930a).

Smith had specific requests with regard to how his collection would be treated in the Columbia Libraries. On December 8, 1930 Smith wrote to C. C. Williamson, the Director of Columbia's Libraries, and Roger Howson Esq, the Librarian of Columbia University:

> I contemplate giving to the Library of Columbia University, in installments from time to time, the major portion of my personal library, excluding books that may

7. Columbia University Receives a Gift

more appropriately go to the Library of Teachers College. This part of my library is very rich in books, manuscripts, portraits, letters, pamphlets, and other material relating to the history of mathematics, medieval life, and the development of books. At present I am prepared to present a considerable part of my orientalia and of my early autograph material. I wish to inquire if this material will be acceptable on the condition that I may retain, or withdraw from the library from time to time, any books or other material thus presented, and to keep the same as long as I may wish and without incurring any risk for loss or damage, this condition applying to future as well as to present gifts. I wish also to place the further condition upon the gifts, that so far as possible the one in charge of these books be authorized to arrange these loans to me without the formality of requiring me personally to draw the material through the general loan desk or to be held responsible for its return at any special time.

My purpose in the matter of retention of loans is to allow me to keep in my home or office as long as I may wish a certain number of rare or interesting pieces.

Upon receipt of an official confirmation of this understanding, I shall prepare as rapidly as I can for the labeling of the books by the Library authorities and the transfer of such material as I do not need for my work or interests.

When Mr. Plimpton presents his library to the University, I hope that as much of my material as he cares to have with his books may be so placed as part of the Plimpton Library, the bookmarks showing that they are presented by myself. (Smith, 1930b, p. 1-2)

7. Columbia University Receives a Gift

About ten days later, on December 17, 1930 Columbia agreed to Smith's terms. In Williamson's response confirming this, he wrote two versions of the same letter. The first version contained the following paragraph:

> I find it impossible to express adequately our feeling of gratitude and appreciation for such a splendid addition to the book and manuscript resources of the University Library. In the fields in which you have been especially interested it will make the Columbia Library the richest collection to be found in this country, if not in the world. Students and scholars for generations to come will be drawn to Columbia to use the materials which could have been brought together only by a man of your thorough scholarship, broad interests, sound judgment, and bibliographical skills. In due time the Trustees will of course make suitable recognition of your generous gift of a collection which will at once and hereafter take rank as one of the most important ever received by the University. (Williamson, 1930b, p. 1)

This version of the letter, which was sent only to Smith, was not meant to be viewed by Plimpton because it was felt by Howson that "[it] may seem to Mr. Plimpton to be detracting from the value of his [collection]" (Williamson, 1930a, p. 1).

In 1932, Plimpton began donating his collection to Columbia. Smith, Plimpton's main advisor on historical mathematics texts, was a major influence in this decision as well. On September 30, 1929 C. C. Williamson wrote to Columbia University's President Nicholas Murray Butler regarding the acquisition of Plimpton's collection. Plimpton's original proposal for how he wished his collection to be cared for was not exactly what Columbia desired. Williamson stated:

> It seems to me that we must be very cautious about

7. Columbia University Receives a Gift

turning down the proposal he makes. Next to the interest of Professor David Eugene Smith such an arrangement as might grow out of his proposal would seem to me to be the greatest possible influence in persuading Mr. Plimpton to give his library to Columbia. If we reject his offer he will almost certainly decide, as he seems inclined to anyhow, that Columbia does not appreciate his collection and is not interested in it. (Williamson, 1929, p. 1)

At this time Plimpton was busy settling his estate; by 1935, the year before he died, Plimpton had his entire collection formally presented to Columbia. As Smith had hoped, his initial deposit of his magnificent collection started a trend among many notable collectors, including Plimpton, which led to Columbia's Libraries' growth. Smith was such a respected and well-liked man that he was able to continually bring in big donations. He remained at the forefront of the Friends Association until 1936 when he had to resign due to age and failing health. At that time the *Bibliotheca Columbiana* also stopped publication, and only two years later, in 1938, the entire Friends collaboration disbanded (Hyde, 1971).

Bertha M. Frick was assigned as the curator/librarian for the Plimpton, Smith, and Samuel S. Dale Libraries in 1937. The Dale Library consists of a collection of 1,200 books and manuscripts along with 700 pamphlets regarding measurement throughout history and across various cultures. Frick was responsible for the current organization of Smith's collection (Columbia University Libraries, n.d.).

Not every piece in Smith's collection ended up being included in the Columbia Libraries, as evidenced in a letter from Smith to Professor Lao G. Simons of Hunter College, dated April 21, 1931:

> In sorting out my portraits my niece found a considerable number of duplicates in the mathematical field.

> The duplicates are not of any value, and in general they are small pieces. It occurs to me, however, that you may wish in your department for some of these... and I wish you would select such as you wish. After you have made your selection, I shall suggest to Dr. Vera Sanford that she do the same when she is here this coming summer. (Smith, 1931, p. 1)

Thus, part of his collection, has been scattered throughout the world.

As he had promised, Smith continued to transfer his collection to Columbia in installments. His next large donation would be his mathematical instruments. Since Smith had previously loaned his mathematical instrument collection to the New York Museum of Science and Industry, he arranged to have all of those items sent over to Columbia (Smith, 1935). In 1935 he gave his collection of about 275 astronomical and mathematical instruments. This included "an Alexandrian terra cotta zodiacal table through telescopes, abaci, spheres, tally sticks, and the like, which he gathered in the various countries he visited" (McAleer, 1961, p. 19). This donation was widely recognized, and Smith responded by writing an article related to his gift to provide more detailed descriptions of some of his pieces (Smith, 1936b).

Smith's collection now had a new home, and even more publicity being directly associated with Columbia University. Some of the major institutions would refer their inquiries to Smith. A letter dated September 17, 1934 to Smith from A. van Niekerk stated, "recently I wrote the Smithsonian Institute for some information relative to logarithmic tables; in the reply I received from the Librarian of that Institute, the suggestion was made that I should take up the matter with you" (van Niekerk, 1934, p. 1).

Frick helped organize numerous exhibits throughout the years and was a trusted friend to Smith. As would be expected, Smith remained involved with his collection until his death in 1944. Two years after Smith's death, Frick resigned as curator, though many of her original organizational plans for his collection have endured (Columbia University Libraries, n.d.). In response to Smith's death, many newspaper and journal articles were dedicated to him in memoriam. Most of these stated that his collecting was what stood out among his lifelong accomplishments.

7.2 Smith's Collection Today

Exhibitions of Smith's collection continued at Columbia University, public libraries in the New York area, and the American Museum of Natural History, to name a few, through the years after it was donated to the University. Once at Columbia, Smith's collection was housed in the Rare Book Department in Low Memorial Library in Room 209—a fireproof vault—as he thought necessary at Teachers College and stressed it was very pressing to Dean Russell in 1915 (Smith, 1915, p. 1). As of July 1, 1944, the Smith, Plimpton, Dale, and Typographic Libraries, along with the Special Collections of the West Wing of Low Memorial Library, were merged into the Department of Special Collections. Smith, Plimpton, and Dale's collections were then housed in room 210 of Low Memorial Library. In 1950 these libraries were transferred to Butler Library, where they are today (Columbia University Libraries, n.d.).

The most recent exhibit, "'The Ground of Arts:' Mathematical Instruments and Illustrated Books from the David Eugene Smith Collection," was from December 5, 2002 to February 28, 2003 (Figure 7.2.1). Currently, Smith's collection is available to view on a piece-by-piece basis through the Rare Book and Manuscript Library on the sixth floor of the Butler Library.

108   7. Columbia University Receives a Gift

Figure 7.2.1: Part of Case X at the 2002 Exhibit. Includes nests of weights and money changer's balances. Image courtesy of the Rare Book and Manuscript Library, Columbia University.

7.3 References

Columbia University Libraries. (n.d.). Finding aid for Rare Book and Manuscript Library records, 1917-2006. Rare Book and Manuscript Library, Columbia University, New York.

Hyde, M. C. (1971). History of library friends and the phoenix story of Columbia. *Columbia Library Columns, XX*(3).

McAleer, H. E. (1961). A family portrait of 'U.D.'. *Columbia Library Columns, X*(3).

Smith, D. E. (1915, November 3). [Letter to James E. Russell]. David Eugene Smith Professional Papers, 1860-1945 (Box 41). Rare Book and Manuscript Library, Columbia University, New York.

Smith, D. E. (1930a, September 19). [Letter to Prince Damrong Rajanubhab]. David Eugene Smith Professional Papers, 1860-1945 (Box 48). Rare Book and Manuscript Library, Columbia University, New York.

Smith, D. E. (1930b, December 8). [Letter to C. C. Williamson]. David Eugene Smith Professional Papers, 1860-1945 (Box 54).
Rare Book and Manuscript Library, Columbia University, New York.

Smith, D. E. (1931, April 21). [Letter to Lao G. Simons]. David Eugene Smith Professional Papers, 1860-1945 (Box 48). Rare Book and Manuscript Library, Columbia University, New York.

Smith, D. E. (1935, September 16). [Letter to Bertha M. Frick]. David Eugene Smith Professional Papers, 1860-1945 (Box 48). Rare Book and Manuscript Library, Columbia University, New York.

Smith, D. E. (1936b). The David Eugene Smith gift of historical-mathematical instruments to Columbia University. *Sci-*

ence 83(2143), 79-80.

van Niekerk, A. (1934, September 17). [Letter to David Eugene Smith]. David Eugene Smith Professional Papers, 1860-1945 (Box 52). Rare Book and Manuscript Library, Columbia University, New York.

Williamson, C. C. (1929, September 30). [Letter to David Eugene Smith]. David Eugene Smith Professional Papers, 1860-1945 (Box 54). Rare Book and Manuscript Library, Columbia University, New York.

Williamson, C. C. (1930a, December 15). [Letter to David Eugene Smith]. David Eugene Smith Professional Papers, 1860-1945 (Box 54). Rare Book and Manuscript Library, Columbia University, New York.

Williamson, C. C. (1930b, December 17). [Letter to David Eugene Smith]. David Eugene Smith Professional Papers, 1860-1945 (Box 54). Rare Book and Manuscript Library, Columbia University, New York.

Chapter 8

Conclusion

The importance of incorporating the history of mathematics by way of objects and artifacts was an enduring contribution that David Eugene Smith made to the teaching of mathematics. Smith described it best:

> My hobby is the human side of the cold and austere science of mathematics. I like to see how mathematics has woven itself into human life through millions of years, and to see how millions upon millions of our race solved the problems of this science before they solved the greatest of all the problems of the universe. I like to see in the story of mathematics is the story of the escape of humanity from the prisons of superstition, how through the measurements of mathematics it recognized, as other worlds, the stars which it had once thought to be merely silver points in crystal spheres; how it struggled to learn its number systems, write its figures, and to appreciate the marvels of a mere zero. I like to see its efforts at computing and to watch it as we watch a child count his blocks, his pebbles, or his fingers and toes. (Smith, 1917, p. 2)

8. Conclusion

The objects of mathematical practice through the ages can create a bridge to connect abstract theories with daily reality, and Smith knew the power of it as a teaching tool. It was evident that his collection was a way to draw in a student of any age, and Smith invited everyone he encountered to join him in his quest for discovering mathematics through its history. The items he devoted a lifetime to assembling were not merely a collection of beautiful and rare pieces of mathematics history—but an extension of Smith as a person. Students of Smith were not shy in their admiration; for example, upon his retirement they commissioned a portrait of him, and mathematical clubs were named after him (Williams, 1935). This portrait hangs today in the Mathematics Education Program office of Teachers College. Recently some copies of portraits of mathematicians from Smith's collection were framed and placed on the walls of the office, somewhat similar to how it was decorated during Smith's time (Figure 8.0.1). A statement by Clifford B. Upton made on July 30, 1930 is as true today as it was almost one hundred years ago, "I had not realized before that the Smith Collection contained so many individual items. It is certainly a remarkable collection. I wish we had it at Teachers College" (Upton, 1930, p. 1).

As Smith's travel journals reveal, the life of a collector in the early twentieth century was far from easy. The process of collecting involves not only knowledge of the items sought but also a sense of adventure—not to mention the strength to barter. In some cases, Smith had to be fearless to complete an acquisition. It is apparent that Smith had an intense passion for collecting and the vision to share his treasures with the world. Throughout his career, Smith had a deep passion for collecting and sincerely believed that access to mathematical objects was instrumental to the proper teaching of the history of mathematics and to mathematics education in general.

8. Conclusion

Figure 8.0.1: Department of Mathematics Office, circa 1910, with prints of celebrated mathematicians on the back wall. Image is provided courtesy of the Gottesman Libraries at Teachers College, Columbia University.

8. Conclusion

Through Smith's dedication and vision for the Educational Museum, he was able to provide for the students, educators, and public a remarkable place to come view current and historical pieces of mathematics education. When the Educational Museum of Teachers College had to close, his collection was able to be viewed by the public through George A. Plimpton's Permanent Educational Exhibit, George F. Kunz's Museums of the Peaceful Arts, and the New York Museum of Science and Industry. Along with Smith's own words regarding his travels, the collection itself is described using Smith's narratives from a talk given to the Hobby Club of New York City. These accounts provide a window into the magnificent and broad scope of the collection as it was viewed by Smith's contemporaries.

Smith's connection with Columbia University's Libraries grew beyond mathematics education, as he was the founder of the Friends of the Library of Columbia University which began the trend of donating exceptional collections, including Smith and Plimpton's, to the University. The collection is now located in the Rare Book and Manuscript Library of Columbia University located on the sixth floor of Butler Library.

Smith has been recognized for his many accomplishments and contributions to mathematics education. He was a mathematics professor, historian, author, librarian, editor, and collector. Serving as the librarian of the American Mathematical Society from 1902–1920 gave Smith great satisfaction, as did being an editor for their *Bulletin*. One of his major contributions to mathematics education was the organization and leadership of the International Commission on the Teaching of Mathematics from 1908-1912. He was the president of the Mathematics Association of America in 1920, where he also was an editor for the *American Mathematical Monthly*. Smith founded the History of Science Society in 1924, and along with Jekuthiel Ginsburg established the *Scripta Math-*

8. Conclusion

ematica journal in 1932. Besides those remarkable achievements, Smith was also a full-time tenured professor, wrote numerous textbooks and history of mathematics books, and travelled around the world collecting. It is hard to imagine that one man did all this, and yet, he did it brilliantly.

The objects in Smith's collection provide a direct link to the past that can ignite inspiration by simply glancing at an item. It would have been incredible to have known him, to have heard his stories in person, and to have visited his home to view his treasures—as was a common invitation for anyone he knew—would have been quite an honor.

Bertha Frick was very close to him and his collection; she understood that side of Smith quite well and described him:

> One of the clearest cut pictures of Dr. Smith is of him seated in his spacious living room surrounded by his dearest art objects—rugs from Persia and Afghanistan on the floor, Russian ikons and Chinese sketches on the walls, bowls from Java, Buddhas from Burma, delicate glass from Syria in cases and on tables where he could see and touch with loving hands each of these treasures. As he often said, 'I love to sit here and let my eyes wander, wherever they fall I live again my experiences in finding it, down a river infested by crocodiles, seated cross-legged in the mud hut of a native chief, wandering among the ruins of an ancient Buddhist monastery, ghostly in the moonlight.' How he loved to tell these stories to an interested visitor! And his guest traveled with him to these far-off places. (Frick, 1944, p. 2)

This book has taken the reader on that journey, as Smith would have described it. Tracing his collection through its many adventures around the world, its journey through numerous museums, and the powerful inspiration it conjures for research opportunities. Although this amazing collection has had many homes and David Eugene Smith has passed on almost seventy years ago, his collection remains at Columbia University while his passion for it will live on in the work of future teachers and historians of mathematics.

8.1 References

Frick, B. M. (1944, August 11). Dr. D. E. Smith, mathematician, world traveler. *Cortland Democrat.*

Smith, D. E. (1917). [Notes on talk, Mirabilia Mathematica, given to Hobby Club]. David Eugene Smith Professional Papers, 1860-1945 (Box 88). Rare Book and Manuscript Library, Columbia

Upton, C. B. (1930, July 30). [Letter to David Eugene Smith]. David Eugene Smith Professional Papers, 1860-1945 (Box 52). Rare Book and Manuscript Library, Columbia University, New York.

Williams, J. (1935, January 17). [Letter to David Eugene Smith]. David Eugene Smith Professional Papers, 1860-1945 (Box 54). Rare Book and Manuscript Library, Columbia University, New York.

Appendix A

David Eugene Smith's "Illustrations for Lectures on the History of Mathematics" Series I and II (1907)

The following transcriptions of Series I and II are from two pamphlets from the Educational Museum of Teachers College. These were considered "catalogs" for ordering slides of items in the collection at the Educational Museum. These documents were printed in 1907 and distribution of slides stopped as of 1914. They are reproduced under permission from the Rare Book and Manuscript Library, Columbia University.

A. Illustrations for Lectures

Transcription of Series I

In response to requests for the use of stereopticon slides illustrating the history of mathematics, the Educational Museum takes pleasure in announcing that it has made arrangements for supplying this material to schools and colleges. Since the demand thus far has been greatest for illustrations showing the development of arithmetic, the following list relates chiefly to that subject. If, however, there should be a demand for slides illustrating the growth of algebra, geometry, trigonometry, analytic geometry, and the calculus, these can also be supplied. The large collections of instruments, rare books, portraits, manuscripts, and photographs of material in foreign museums and libraries of the University and Teachers College, as well as those in George A. Plimpton, Esq., and Professor David Eugene Smith, afford opportunity for the preparation of illustrations in nearly every branch of the subject. Brief accounts of these collections will be sent upon request.

Should a sufficient demand be expressed the Museum will consider the question of making similarly available the resources of other departments of the College.

The slides will be furnished only to schools and colleges, or to those who give courses in such institutions. Since the price represents merely the cost to the Museum, no discount can be allowed, whatever the number purchased. The arrangement with the photographer requires that no order for less than twenty-five (25) slides shall be accepted. The price is $10 for twenty-five slides, and 40 cents each for any number in excess.

A. Illustrations for Lectures 119

Since many of the slides are not kept on hand, there will be a delay of two or three weeks in filling any order. Orders should be addressed to

The Educational Museum
Teachers College
Columbia University
New York, NY

EGYPTIAN

1. First trace of Egyptian mathematics, a pottery inscription of the first dynasty.

2. Page from the Ahmes papyrus, c. 1700 B.C., the oldest extant textbook on mathematics.

3. Page from the Akhmim papyrus, possibly of the 8th century A.D., showing the same primitive treatment of fractions as in Ahmes.

OUR NUMERALS

4. Nana Ghat inscriptions. See Encyclopedia Britannica, under "Numeral."

5. Same in detail.

6. Page from a MS. of Boethius of 1286, showing forms of numerals.

7. From a MS. of Rollandus of 1420, showing forms of numerals.

8. From a MS. of Sacrobasco's "Algorismus" of 1444, showing forms of numerals.

9. From a MS. of the 15th century, showing forms of numerals.

10. Table from Treutlein's "Zahlzeichen," showing the changes in the numeral system from ancient to modern times.

NUMBER NAMES

11. Page from Borghi's arithmetic of 1488, showing one of the early uses of "million" in print.

PRACTICAL USE OF ROMAN NUMERALS

12. The Roman numerals in practical use in 1514, from Kohel's arithmetic.

13. The same, showing the curious use of Roman numerals with Arab fraction forms.

ABACUS, OR LINE RECKONING

14. Page from an "Algorithmus Linealis" of c. 1490, showing the reckoning with counters.

15. Page from Licht's "Algorismus" of 1501, showing addition by means of counters.

16. Picture from the "Margarita Philosophica" (1503), showing the old (counter) and new (algorism) reckoning.

17. Title page of one of Adam Riese's arithmetics (1538), showing merchants reckoning "on the line."

18. The same, from Gemma Frisius (1565 edition).

19. The same, from Recorde's "Ground of Artes" (1558 edition). (Riese, Gemma Frisius, and Recorde were the most popular arithmeticians of their time.)

20. Addition by counters, from Recorde's "Ground of Artes" (1558 edition).

21. Chinese swanpan, Russian tschotü, and Korean rods, the modern relics of the counter reckoning.

MODERN MECHANICAL COMPUTATION

22. Machines for adding, multiplying, and dividing.

FINGER RECKONING

23. The ancient finger reckoning as illustrated in the "Abacus" of Aventinus (1532).

24. The same, from Recorde's "Ground of Artes" (1558 edition).

SYMBOLS

25. Early use of the symbol =, before it was used or equality, from an anonymous MS. of c. 1450.

26. Earliest use of a decimal point (Pellos, 1492), about a century before decimal fractions were understood.

27. First printed page containing the signs + and − (Widman, 1489), as symbols of excess and deficiency.

28. Early use of the same as algebraic symbols (Stifel, 1545).

29. Symbols of addition and subtraction from Curtius (1619), with curious processes of multiplication.

FUNDAMENTAL OPERATIONS

30. Numeration by the catechism method, from Willichius (1540).

31. Addition, from Hylles (1579), showing curious rhyming rule, and catechism method of teaching arithmetic.

32. Subtraction (substraction), from Baker's "Well Spring of Sciences" (1580).

33. Multiplication. Elaborate specimen of the *gelosia* method, from a MS. of c. 1450.

34. Multiplication *per scachiero* and *per quadrato*, from a 15th-century MS.

35. Multiplication per gelosia, from Feliciano's arithmetic of 1545.

36. The old complementary multiplication from Huswirt's "Enchiridion" of 1501.

37. Multiplication, with curious illustration, from a student's MS. of 1561.

38. Multiplication and division as performed in the first printed arithmetic (Treviso, 1478).

39. Division by the galley method and multiplication by the common (*scachiero*) plan, from a MS. of the 16th century.

40. Division by the galley method, showing the galley, from a Venetian MS. of c. 1550.

41. Division by the galley method, from a MS. of c. 1600.

42. A very early specimen of the modern form of division, from a MS. of c. 1450.

A. Illustrations for Lectures 123

43. The first printed example of our modern (*a danda*) form of division, from Calandri (1491).
44. Modern division, with curious forms of the numerals, from a MS. of c. 1550.
45. Division of fractions, with curious symbols and proofs, from a MS. of 1545.
46. Cube root by the galley method, from the first arithmetic printed in England (Tonstall's "De Arte Supputandi," 1522).

MEDIAEVAL PROPORTION

47. From a MS. of Boethius written in 1288, giving the arithmetical, geometric, and harmonic proportion.
48. The same, with musical proportion from the first printed edition of Boethius (1488).
49. Proportion as the Rule of Three. Examples from Fisher (1775 edition).

SLATE AND BLACKBOARD

50. The first printed mention of a slate (Prosdocimo de Beldamandi, 1488).
51. Curious illustration from a MS. of Sacrobosco, written in 1444, showing master teaching the new numerals.
52. Title page of Boschenstein's arithmetic of 1514, showing merchants using the blackboard.

THE ANCIENT SCHOOL

53. A class in arithmetic in the Middle Ages, from an old engraving.

54. A mediaeval school, from an old engraving.

55. The seven liberal arts, from an old engraving.

56. The sciences illustrated (Arithmetic with the counters), from an old engraving.

THEORY OF NUMBERS

57. From a MS. of Boethius of 1286, showing figurate numbers.

58. From the first printed edition of Boethius (1488), showing other figurate numbers.

59. From an anonymous chapter on Rithmimachia (1496), showing this famous mediaeval number game.

60. The first printed Magic Square, from Dürer's "Melancholia."

TOPICS STUDIED

61. Title page of Paciuolo's great work of 1494 (1523 edition), giving a list of the important topics.

62. The problem of the Venetian clock, from Kobel's arithmetic (1540 edition).

63. Old treatment of Partnership, from Masterson's arithmetic of 1592.

64. Barter, from Daboll's arithmetic (4th edition).

A. Illustrations for Lectures 125

65. Early American problems from Pike (1788).

66. Problems of the Civil War, from Johnson's arithmetic (Raleigh, N. C., 1864).

THE ILLUSTRATING OF ARITHMETICS

67. The problems of the jealous husbands, and the jugs, from a 14th-century MS.

68. The chessboard problem of the grains of wheat, from a 14th-century MS.

69. From Sacrobosco's "Sphaera" (Venice, 1488), showing mediaeval theory of the apparent rotundity of the sea.

70. From the first printed arithmetic having illustrations. Calandri's book of 1491.

71. From Widman's arithmetic of 1489, showing illustration in exchange.

72. From Kobel's arithmetic (1544 edition), showing one of the problems of the couriers.

73. From the same, showing the problem of the market women.

74. Humorous illustrations from Crowquill's arithmetic (1848).

FAMOUS ARITHMETICS

75. Last page of the first printed arithmetic (Treviso, 1478).

76. First page of the rare "Ars Numerandi" (c. 1485, but possibly as early as the Treviso).

77. Last page of the first German arithmetic (1482).

126   A. Illustrations for Lectures

78. Last page of the second German arithmetic (1483).

79. Last page of Calandri's arithmetic (1491).

80. First page of Paciuolo's great treatise of 1494 (1523 edition).

81. First page of the part on arithmetic in Capella's work (1499).

82. Last page (colophon) of Tzwifel's arithmetic (1507).

83. Title page of Bonini's arithmetic (1517), with De Morgan's autograph.

84. Title page of Feliciano's arithmetic of 1526 (1536 edition).

ALGEBRA

85. From the Rollandus MS. (c. 1420), showing the names for the powers of the unknown, and a multiplication table of such powers.

86. Introduction to algebra, from an Italian MS. of c. 1450.

87. From the same MS., with a reference to the work of Leonardo of Pisa.

88. From the MS. of Scheubel's algebra, 16th century, showing his symbolism for surds.

89. The first printed solution of the cubic equation, Cardan's "Ars Magna" (1545).

90. From Masterson's work of 1592, showing the Renaissance symbolism for the unknowns.

91. From a MS. of c. 1620, showing the extraction of the square root of a binomial surd.

GEOMETRY

92. Page from the Campanus translation of Euclid, showing the Pythagorean theorem. Original MS. of c. 1260, in the Plimpton Library.

93. Page from a later Campanus MS. of Euclid, c. 1288.

94. Illustration of *Geometria*, with quadrans, from the "Margarita Philosophica" (1503).

95. From Foeniseca's "Opera" (1515), showing the construction of the Platonic bodies.

96. From Recorde's "Castle of Knowledge" (1596 edition), showing the geocentric idea of the universe.

97. From the "Protomathesis" of Finaeus (1532), showing the two forms of the quadrans.

98. From Beutel's "Lustgarten" (1600), showing the use of primitive instruments in mensuration.

GREAT MATHEMATICIANS

99. Pythagoras, from Calendri's arithmetic (1491).

100. Euclid, from an old engraving.

101. Ptolemy and Boethius, from a drawing by Raphael.

102. Claude Ptolemy, from the "Margarita Philosophica."

103. Leonardo of Pisa, from an engraving.

104. Adam Riese, the most influential German textbook writer in the 16th century, from an old lithograph.

A. Illustrations for Lectures

105. Gemma Frisius, the most successful writer of a Latin arithmetic in the 16th century, from a contemporary engraving.

106. Clavius, one of the first writers of a practical textbook on algebra, from a contemporary engraving.

107. Cardan, from a contemporary engraving.

108. Tartaglia, from a contemporary engraving.

109. Napier, from a rare lithograph.

110. Bachet de Meziriac, editor of Diophantus, and the first to compile a noteworthy collection of mathematical recreations.

111. Descartes, from an engraving after the Hals painting.

112. Fermat.

113. Pascal.

114. Newton.

115. Leibnitz.

116. Euler.

117. Cocker, the greatest writer of arithmetics in England in the 17th century.

118. Dilworth, Cocker's successor in the 18th century.

119. A collection of autographs, including Hermite, Eucler, Legendre, Monge, Johann Bernoulli, Lagrange, Sylvester, Laplace, and other.

To this list may be added the illustrations in Professor Smith's "Rara Arithmetica" (May, 1907), and these may be ordered by specifying the pages. This work also furnishes descriptive matter for many of the slides mentioned in the above list.

The slides above described are prepared largely from the original works in Mr. Plimpton's library.

The circular of the Department of Mathematics, containing Miss Benedict's article on "Algebraic Symbolism," will be sent to teachers interested in the history of mathematics who will send their names and addresses to the Secretary of Teachers College, Columbia University, New York City.

Transcription of Series II

On account of the great demand for stereopticon slides illustrating the history of mathematics, resulting from the circulation of the first list prepared by the Educational Museum, it has been decided to prepare this supplementary list of later acquisitions. The first circular, containing 119 titles, will be sent on request.

The illustrations here mentioned are prepared chiefly from works in the library of Professor David Eugene Smith, although some are from the collection of George A. Plimpton, Esq., and a few are from other sources.

The slides will be furnished only to schools and colleges, or to those who give courses in such institutions. Since the price represents merely the cost to the Museum, no discount can be allowed, whatever the number purchased. The arrangement with the photographer requires that no order for less than twenty-five (25) slides shall be accepted. The price is $10 for twenty-five slides, and 40 cents each for any number in excess.

A. Illustrations for Lectures

Since many of the slides are not kept on hand, there will be a delay of two or three weeks in filling any order. Orders should be addressed to

> The Educational Museum
> Teachers College
> Columbia University
> New York, NY

MISCELLANEOUS

120. From Tagliente's arithmetic (1515), showing curious forms of multiplication.

121. Form an anonymous MS. (c. 1500), giving the horseshoe nail problem.

122. From Bungus, *Numerorum Mysteria* (1614 edition), showing curious forms of Roman Numerals.

123. First page of a MS. of Luca di Firenze, copied c. 1475, showing interesting forms of our numerals.

124. From a MS. of Sacrobosco's *Sphaera*, copied c. 1475, showing the idea of an eclipse.

125. From Reisch's Margarita Philosophica (1504 edition), showing the temple of learning, with Boethius representing arithmetic and Euclid geometry.

126. Title-page of Reisch's Margarita Philosophica (1504 edition), showing goddesses of arithmetic and geometry.

127. First page of Treviso arithmetic (1478). (see No. 75).

128. Last page of a MS. copied by Eustachius de Feltro (1469). Shows only the forms of a few numerals used at that time.

129. Elaborate multiplication table from a Florentine MS. (c. 1500).

130. From an early anonymous work on trigonometry, showing the quadrans.

131. From the Treviso arithmetic (1478), showing elaborate treatment of fractions. (see Nos. 75, 127.)

132. From Tartaglia's arithmetic (1556), showing the galley form of division.

133. Title-page of Coutereels' Dutch arithmetic (c. 1690 edition), showing a reckoning school.

134. From a German arithmetic (c. 1520), showing the Testament problem.

135. Multiplication table, from an anonymous MS. (c. 1400).

136. Title-page of Werner's arithmetic (1561), showing the list of topics then studied.

137. Multiplication table from the arithmetic of Boethius (1488).

138. Initial-page of chapter on Barter in the 1515 edition of Ortega's arithmetic.

139. Two pages from a 1460 MS. of the arithmetic of Benedetto di Firenze, showing the hound and hare problem.

140. From the 1494 edition of Paciuolo's *Summa*, showing finger symbolism. (See No. 80.)

141. From Calandri's arithmetic (1491), showing two pages of illustrated problems. (See No. 70.)

142. Product tables, from the arithmetic of Alexandre Jean (1637).

A. Illustrations for Lectures 133

143. From a child's primer on arithmetic, anonymous (c. 1820), with curious illustrations.

144. Portrait of Erasmus.

159. Multiplication table from Kobel's arithmetic (1514).

160. From the Papyrus Sailier, with an old Egyptian account.

161. From an Italian arithmetic of c. 1525, showing the testament problem illustrated.

169. Title-page of the first (1494) edition of Paciuolo's *Summa*, showing the topics studied. (See No. 140.)

266. From the Ahmes Papyrus. (See No. 2.)

273. Portrait medallion of Lagrange, by David d'Angers.

274. Portrait medallion of Laplace, by David d'Angers.

275. Portrait medallion of Cauchy, by David d'Angers.

OLD MATHEMATICAL INSTRUMENTS

145 to 150. From the geometry of Finaeus (Paris, 1556), showing the various uses of the quadrans, baculus, speculum, and other instruments.

151 and 152. From the *Libro del Misurar*, by Belli (Venice, 1569), showing curious methods of measuring horizontal distances.

153. Early use of the 'Parallelogrammo' (pantograph). From the *Prattica del Parallelogrammo*, by Scheiner (Bologna, 1653).

154 to 157. From the *Modo di Misurare* of Bartoli (Venice, 1589), showing the uses of the quadrans squadro, baculus, etc.

158. Primitive leveling. Form the geometry of Pomodoro, Rome, 1624.

159-161. See after No. 144.

162 to 168. From *De Quadrante geometrico*, by Cornelius de Judeis (Nurnberg, 1504), showing various uses of the quadrans.

169. See after Nos. 144 and 161.

170 to 185. From the *Apiaria Philosophiae mathematicae* (4th edition, Bologus, 1645), by Maria Bettino, showing curious work in the mensuration of distances and surveying.

186. From a German MS. of 1660, with a drawing of the quadrans.

187. From a German MS. of 1660, showing the use of the quadrans.

276. Astrolabes. (1) Italian, 1509; (2) Arabic.

277. Astrolabes. Italian, 1450 and 1558.

278. Sector compasses. Renaissance period.

188 to 199. From Bion's work on *Instrumens de Mathematique* (La Haye, 1723), showing various mathematical instruments and their uses.

200 to 204. From the *Nova Fabricandi Horaria* of Johannes Paulus Gallucius (Venice, 1596), showing various forms of dials.

205. From *L'uso della squadro mobile* by Ottavio Fabri (Venice, 1598) showing the quadrans.

206. From *Del Modo di Misurare* by Bartoli (Venice, 1589), showing the astrolabe in mensuration work.

A. Illustrations for Lectures 135

207. From Forestani's *Arithmetica* (Venice, 1602), showing unusual method of measuring heights.

208-265. See next page.

267. From Alessandro Capra's *Geometria* (Cremona, 1673), showing curious work in leveling.

268. From Fiammelli's La Riga (Rome, 1605), showing the surveyor's riga.

269. The groma used by the Roman and Etruscan surveyors. Drawing from an ancient monument.

270. The Lituus or Augur's staff as used by the Roman surveyors. Drawing from an ancient monument.

271. Monument of Lucius Faustus, a Roman surveyor, showing ancient surveying instrument.

272. Monument of M. L. Macedonus, a Roman surveyor, showing ancient levels.

MODERN MECHANICAL CALCULATION

208. The Comptometer.

209. Principle of the Comptometer.

210. Multiplication on the Comptometer.

211. Type of Comptometer work.

212. The Mechanical Accountant.

213. Beach Adding Machine.

214. The Calcumeter.

215. The Gem Adding Machine.

216. The Universal Adding Machine.

217. The Burroughs Adding Machine.

218. The Standard Adding Machine.

219. The Comptograph.

220. Split key-board of the Burroughs machine.

221. Statement of an account made on a listing machine.

222. Statement of an account made on a listing machine.

223. Invoice made on a listing machine.

224. Burroughs Fractional Machine.

225. Deposit slip made on a listing machine.

226. Trial balance made on a listing machine.

227. Tax accounts made on a listing machine.

228. The Ellis Adding Typewriter, and the National Typewriter Adding Machine.

229. The Arithmograph.

230. The Elliott-Fisher Billing and Adding Machine.

231. Statement of account made on the adding typewriter.

232. Sales sheet made on the adding typewriter.

233. Bank collection letter made on the adding typewriter.

234. Check writing done on the adding typewriter.

235. The National Cash Register.

236. Printed sales-strip made by the Cash Register.

237. The National Cash Register for department stores.

238. Sales-slip printed by the department store Cash Register.

239. The Brandt Change-Making Machine.

240. The Hollerith Electric Tabulating and Adding Machine.

241. Card used in the Hollerith machine.

242. Punch used with Hollerith machine.

243. The Saxonia Reckoning Machine.

244. Principle of Thomas Arithmometer.

245. The Autarith.

246. The Brunsviga Calculating Machine.
     The Triumphator Calculating Machine.

247. The Millionaire Calculating Machine.

248. Manufacturing cost sheet computed by the multiplying machine.

249. Least square solution made by the multiplying machine.

250. The Slide Rule.

251. The Thacher Slide Rule.
     The Fuller Slide Rule.

252. Sperry's Pocket Calculator.
     The Charpentier Calculator.
     The Boucher Calculator.

DEVELOPMENT OF CALCULUS AND ANALYTICS

253. From the Paris (1740) edition of Newton's Fluxions showing symbolism.

254. From Cavalieri's *Geometria Indivisibilibus* (Bologna, 1653), showing a geometric figure.

255. From Haye's *Treatise of Fluxions* (London, 1764), the first work in English on calculus, showing Newtonian symbolism.

256. From Wallis's *Tractatus de Algebra* (Oxford, 1685, 1693 edition), showing first geometric treatment of complex number.

257. From Descarte's *La Geometrie* (1637, 1705 edition), showing first steps in analytics.

258. From Viviani's *De Max. et Minimis* (Florence, 1659), showing early progress towards calculus.

259. From the *Elementa Conica* of Apollonius (Rome, 1679), showing the nature of the definitions.

260. The *Opera* of Archimedes (Rome, 1679), showing first pages *De quadratura parabola*.

261. From the works of Pappus (Bologna, 1660), showing his ratio definition of conics.

262. From Newton's *Arithmetica Universalis* (Leyden edition of 1732), showing symbolism.

263. From Euler's calculus (St. Petersburg, 1708), showing symbolism.

264. A mathematical MS. of Newton in Professor Smith's collection.

A. Illustrations for Lectures

265. A mathematical MS. of Leibnitz in Professor Smith's collection.

266. See after Nos. 144 and 169.

267-272. See after No. 207.

273-275. See after Nos. 144 and 266.

276-278. See after No. 187.

Appendix B

Teachers College's Set of Lantern Slides from the "Illustrations for Lectures on the History of Mathematics"

The following are digitized versions of the slides currently available at the Mathematics Program at Teachers College. On each page, the larger image is the specially scanned version of the slide to display the image portrayed on the lantern slide. The smaller image is a scanned version of the lantern slide in its original state. Each slide is approximately 3"x4". The description at the bottom of the page is taken from the "Illustrations for Lectures on the History of Mathematics" publication. In most cases, this same description is handwritten directly on the slide. Although some use a double numbering system, the slides are organized according to the 1907 publication's ordering. Several of the slides in the set are not from that publication, and thus, are not specifically identified. These most likely came from items in George A. Plimpton's collection.

142   B. Lantern Slides

10

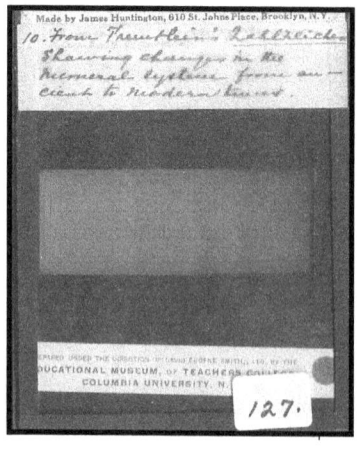

Table from Treutlein's "Zahlzelchen," showing the changes in the numeral system from ancient to modern times.

B. Lantern Slides

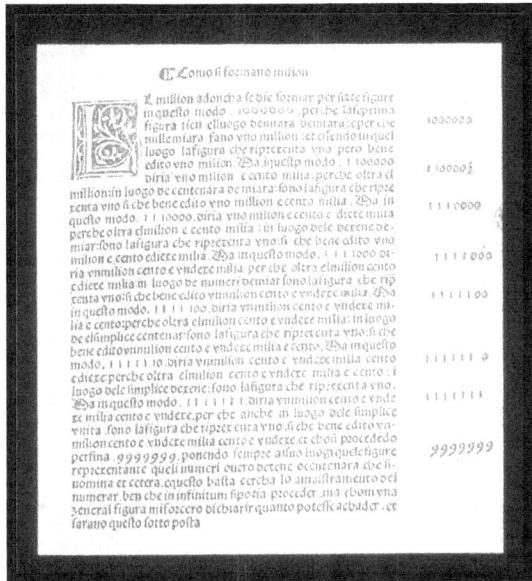

Page from Borghi's arithmetic of 1488, showing one of the early uses of "million" in print.

16

Picture from the "Margarita Philosophica" (1503), showing the old (counter) and new (algorism) reckoning.

B. Lantern Slides

17

Title page of one of Adam Riese's arithmetics (1538), showing merchants reckoning "on the line."

19

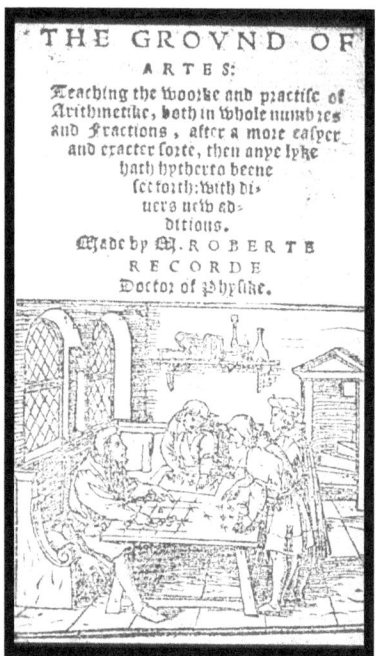

The same, from Recorde's "Ground of Artes" (1558 edition).

B. Lantern Slides 147

21

Chinese swanpan, Russian tschotü, and Korean rods, the modern relics of the counter reckoning.

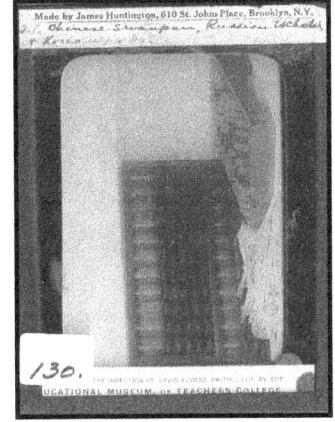

22

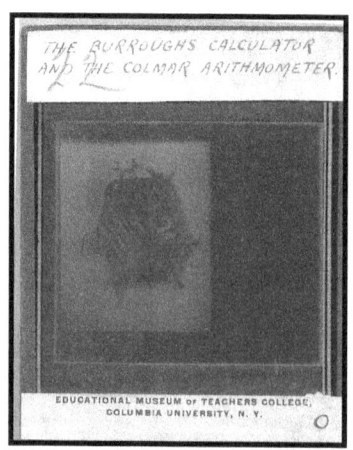

Machines for adding,
multiplying, and dividing.

B. Lantern Slides 149

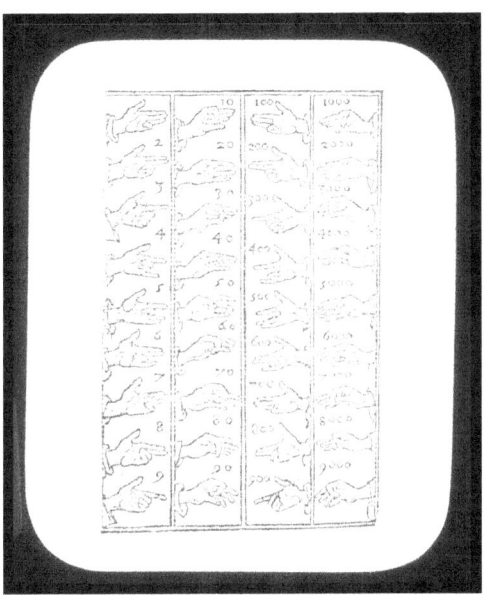

24

[The ancient finger reckoning as illustrated] from Recorde's "Ground of Artes" (1558 edition).

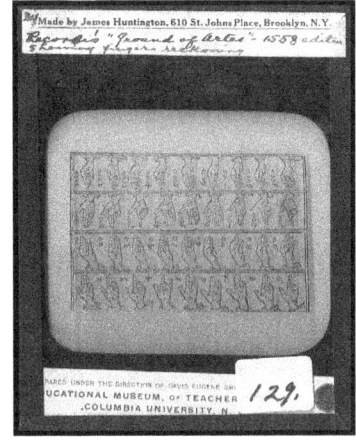

150 B. Lantern Slides

26

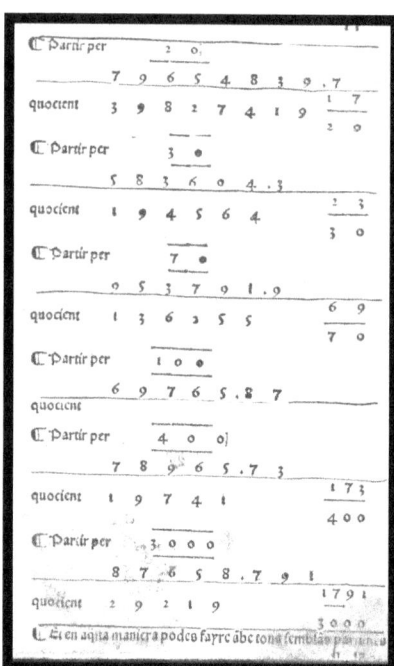

Earliest use of a decimal point (Pellos, 1492), about a century before decimal fractions were understood.

B. Lantern Slides

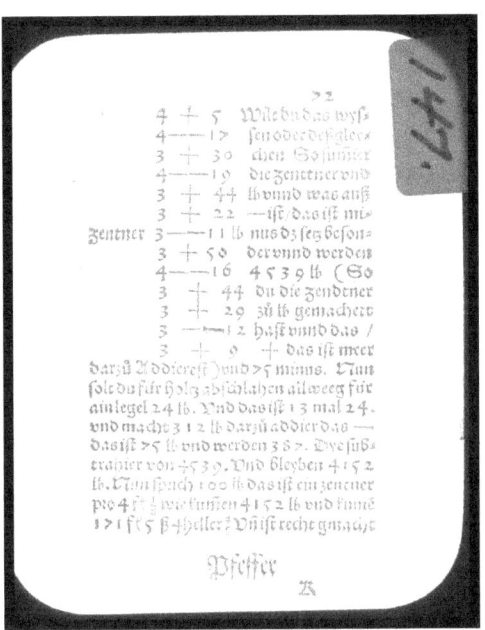

27

First printed page containing the signs + and − (Widman, 1489), as symbols of excess and deficiency.

29

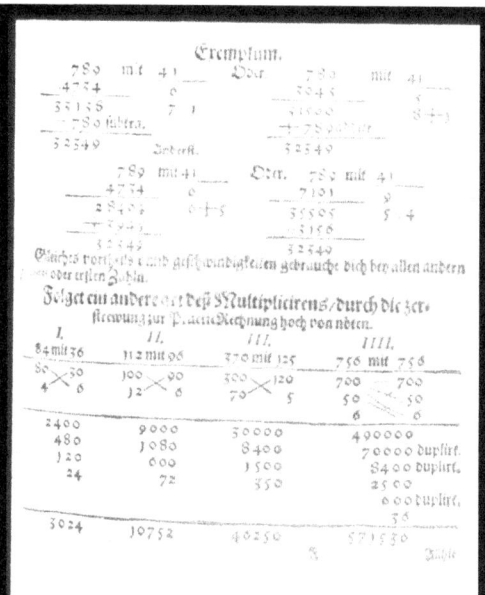

Symbols of addition and subtraction from Curtius (1619), with curious processes of multiplication.

B. Lantern Slides 153

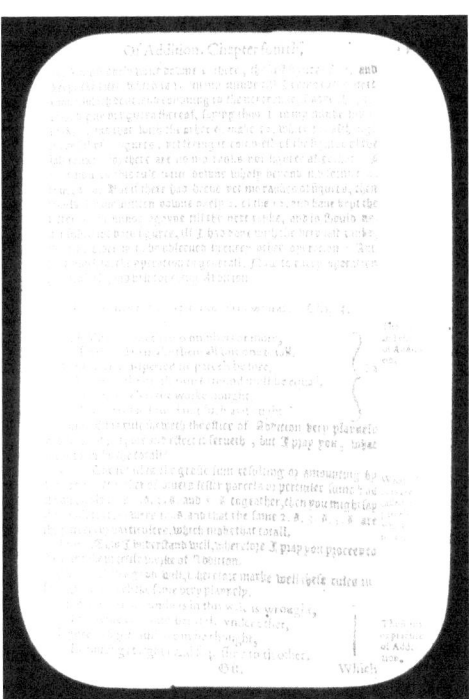

31

Addition, from Hylles (1579),
showing curious rhyming rule,
and catechism method of
teaching arithmetic.

32

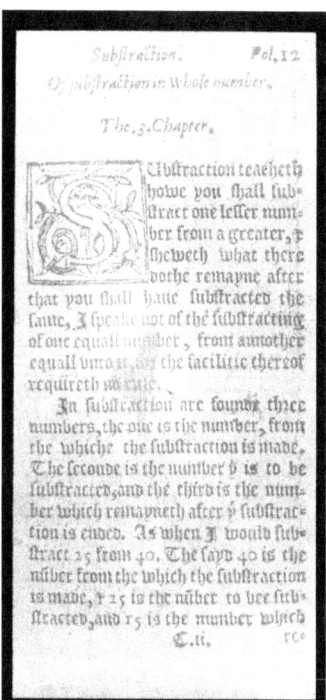

Subtraction (substraction), from Baker's "Well Spring of Sciences" (1580).

B. Lantern Slides 155

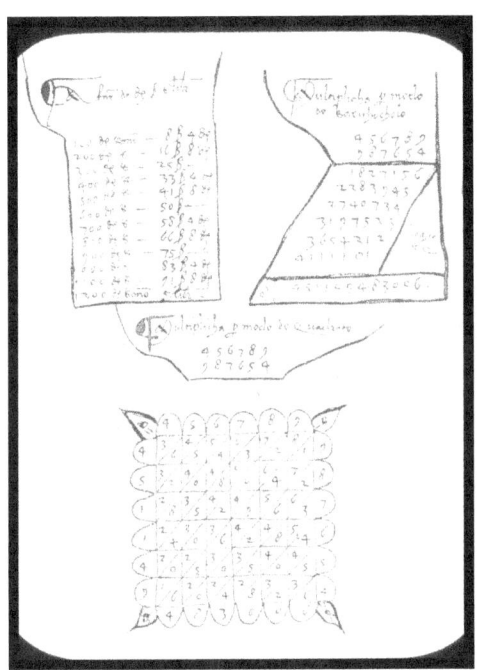

34

Multiplication *per scachiero* and *per quadrato*, from a 15th-century MS.

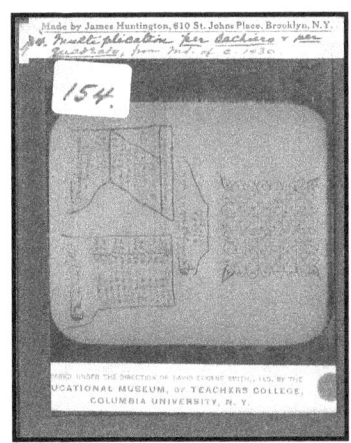

36

The old complementary multiplication from Huswirt's "Enchiridion" of 1501.

B. Lantern Slides 157

39

Division by the galley method and multiplication by the common (*scachiero*) plan, from a MS. of the 16th century.

43

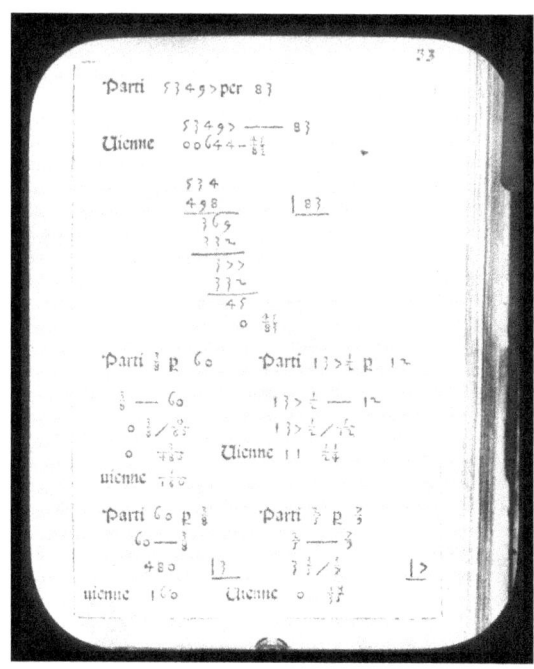

The first printed example of our modern (*a danda*) form of division, from Calandri (1491).

B. Lantern Slides 159

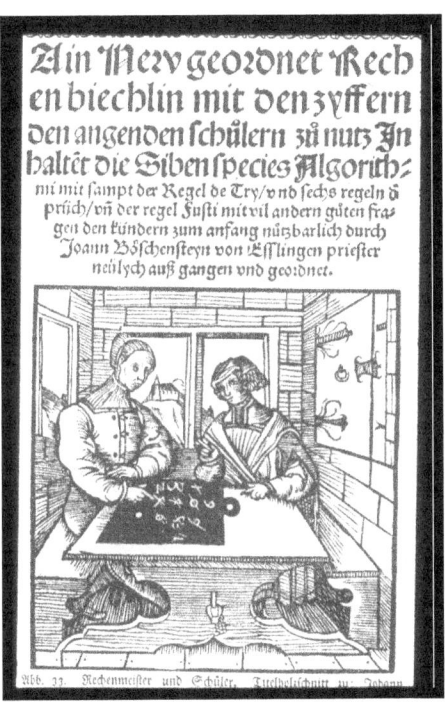

52

Title page of Boschenstein's
arithmetic of 1514, showing
merchants using the
blackboard.

53

A class in arithmetic in the Middle Ages, from an old engraving.

B. Lantern Slides 161

55

The seven liberal arts, from an old engraving.

58

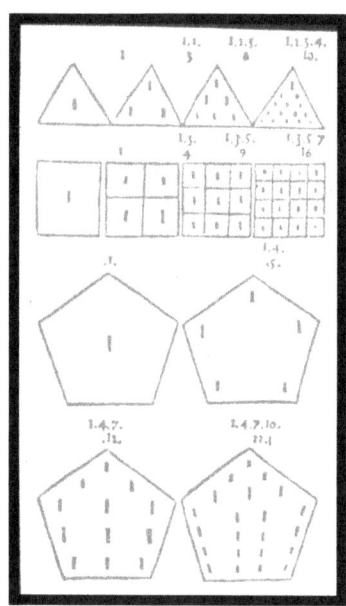

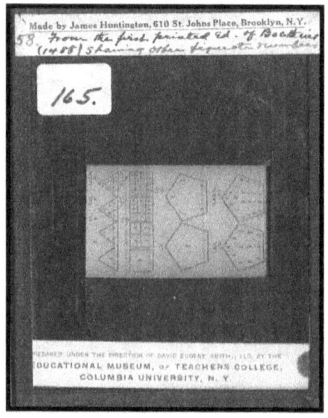

From the first printed edition of Boethius (1488), showing other figurate numbers.

B. Lantern Slides

59

From an anonymous chapter on Rithmimachia (1496), showing this famous mediaeval number game.

60

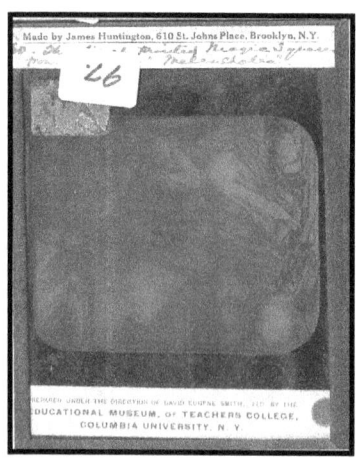

The first printed Magic Square, from Dürer's "Melancholia."

B. Lantern Slides

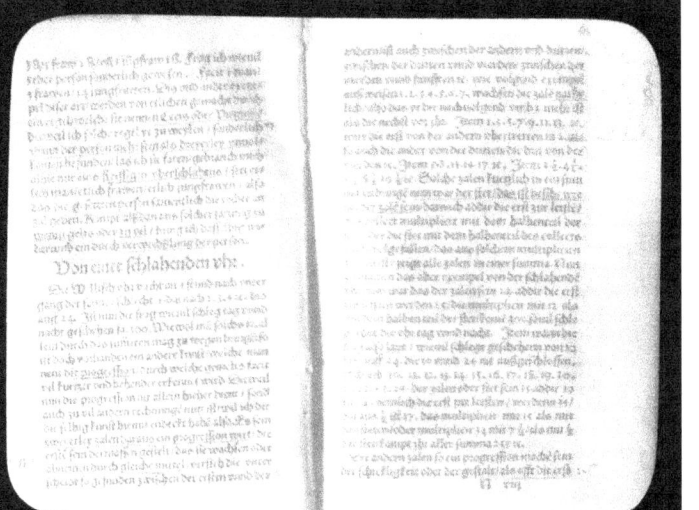

62

The problem of the Venetian clock, from Kobel's arithmetic (1540 edition).

63

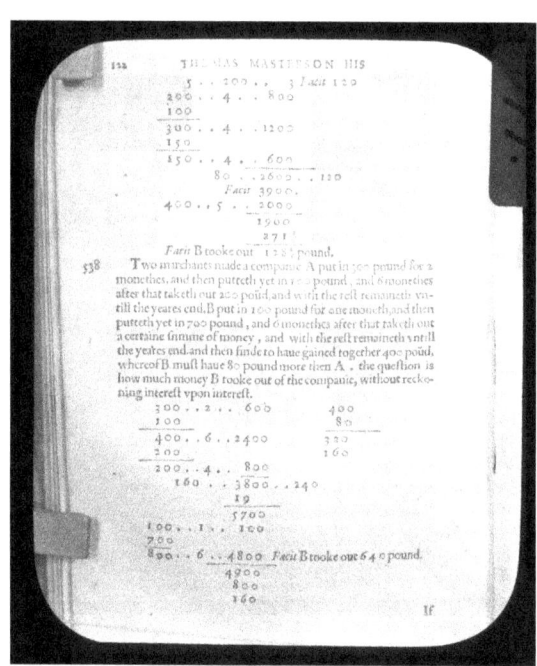

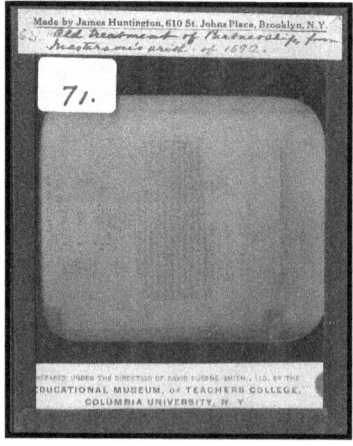

Old treatment of Partnership, from Masterson's arithmetic of 1592.

B. Lantern Slides

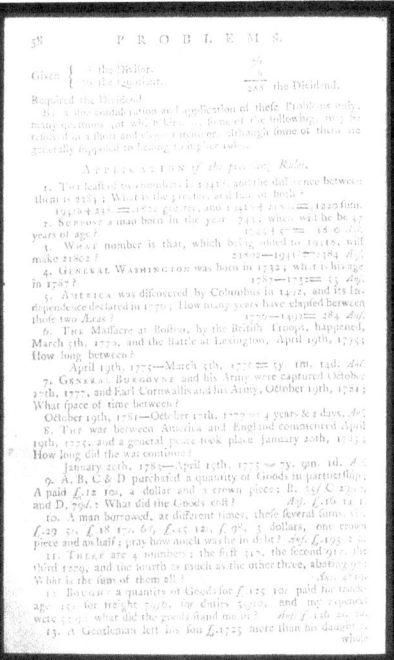

65

Early American problems from Pike (1788).

B. Lantern Slides

74

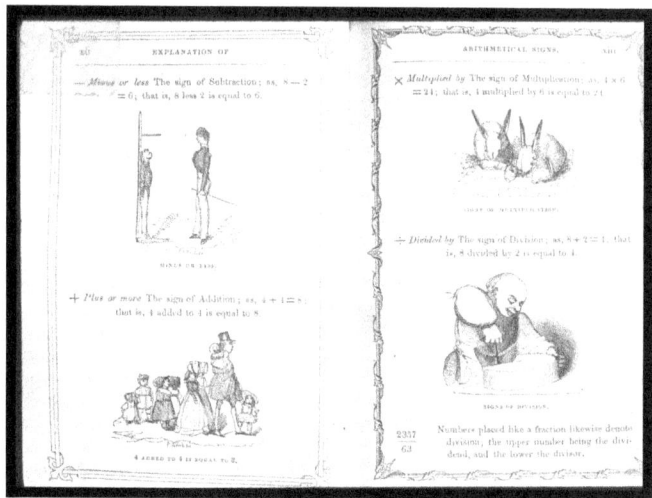

Humorous illustrations from Crowquill's arithmetic (1848).

B. Lantern Slides

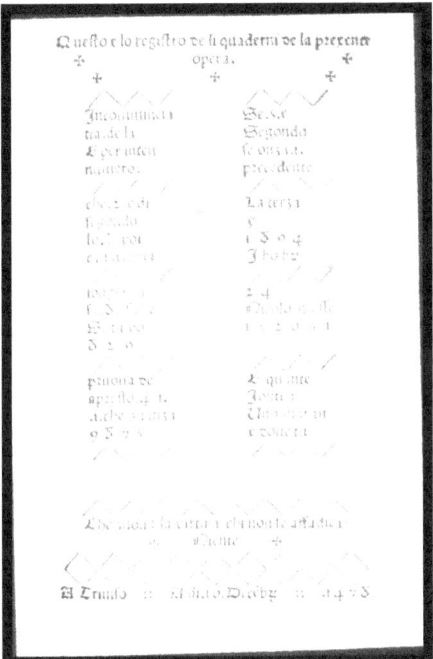

75

Last page of the first printed arithmetic (Treviso, 1478).

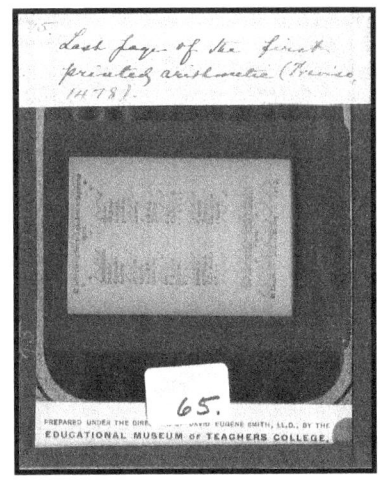

76

First page of the rare "Ars Numerandi" (c. 1485, but possibly as early as the Treviso).

B. Lantern Slides

85

From the Rollandus MS. (c. 1420), showing the names for the powers of the unknown, and a multiplication table of such powers.

86

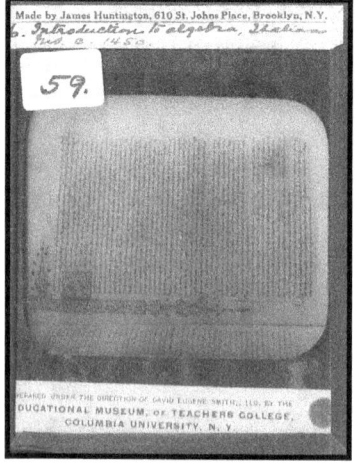

Introduction to algebra, from an Italian MS. of c. 1450.

B. Lantern Slides 173

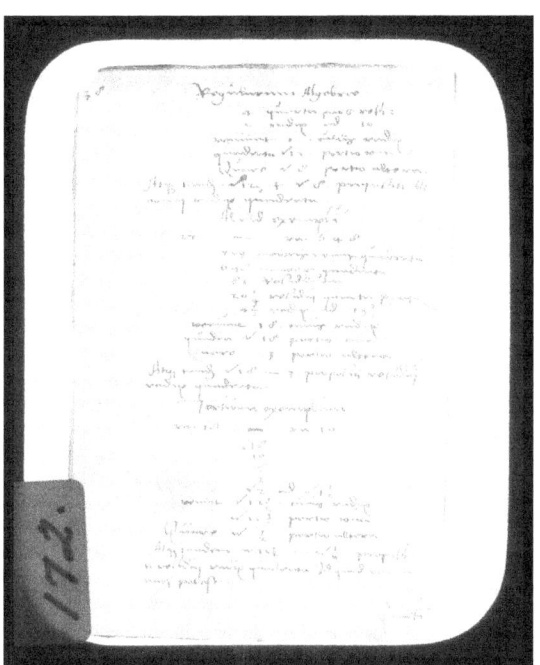

88

From the MS. of Scheubel's algebra, 16th century, showing his symbolism for surds.

89

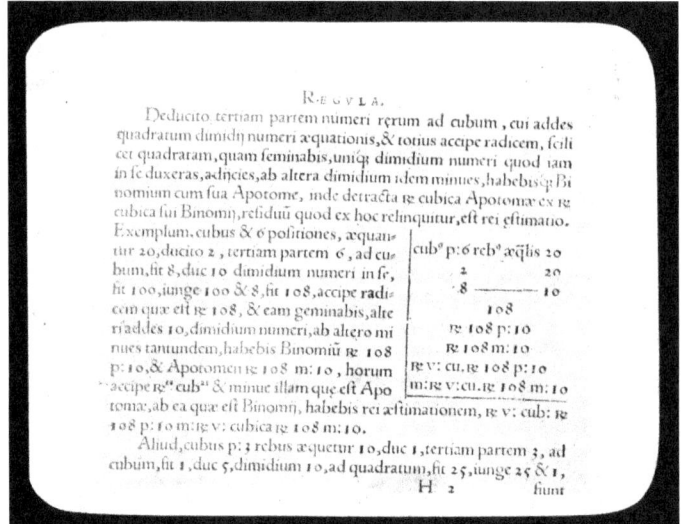

The first printed solution of the cubic equation, Cardan's "Ars Magna" (1545).

B. Lantern Slides

90

From Masterson's work of 1592, showing the Renaissance symbolism for the unknowns.

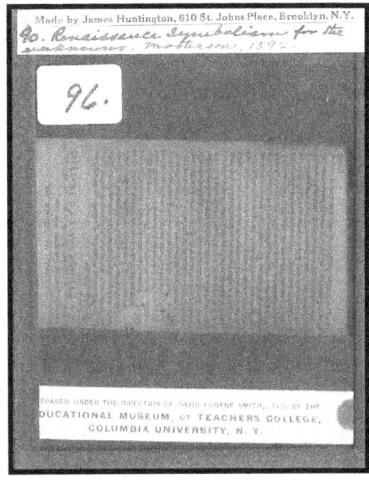

B. Lantern Slides

91

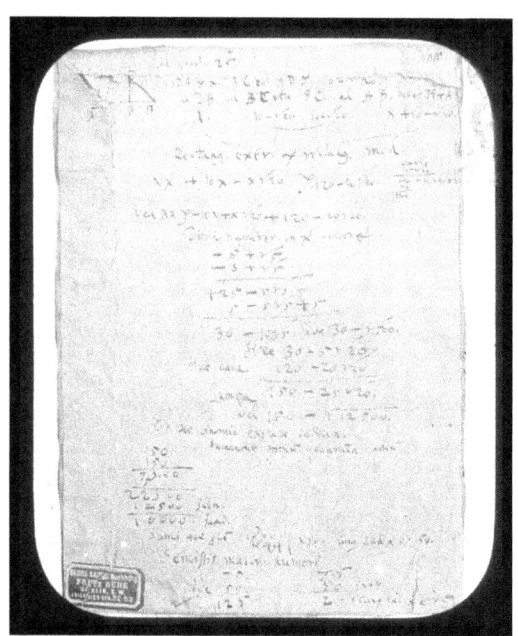

From a MS. of c. 1620, showing the extraction of the square root of a binomial surd.

B. Lantern Slides 177

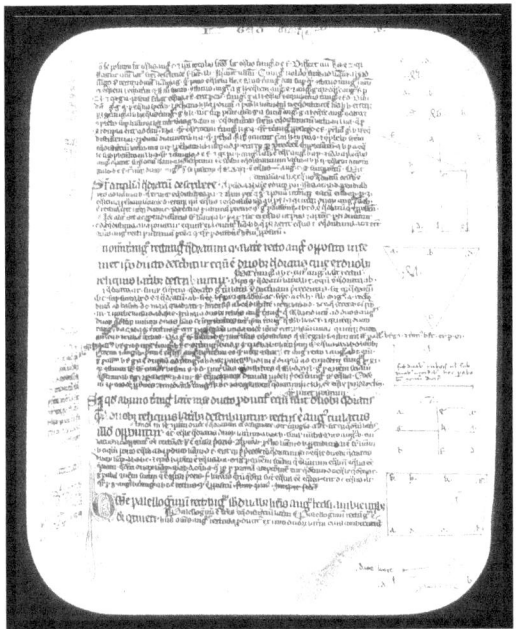

92

Page from the Campanus translation of Euclid, showing the Pythagorean theorem. Original MS. of c. 1260, in the Plimpton Library.

94

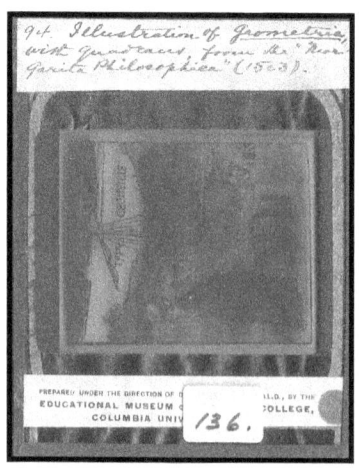

Illustration of *Geometria*, with quadrans, from the "Margarita Philosophica" (1503).

B. Lantern Slides 179

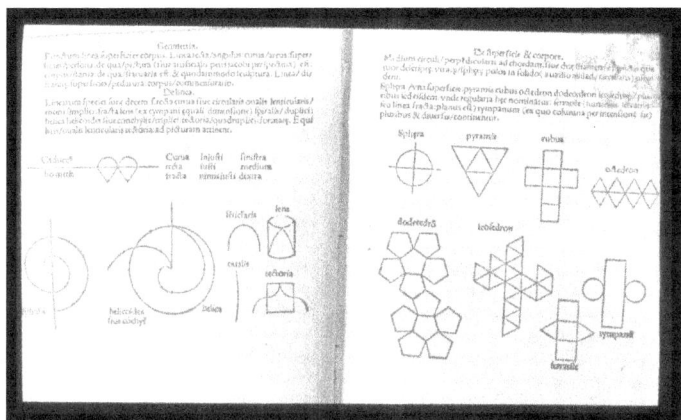

95

From Foeniseca's "Opera" (1515), showing the construction of the Platonic bodies.

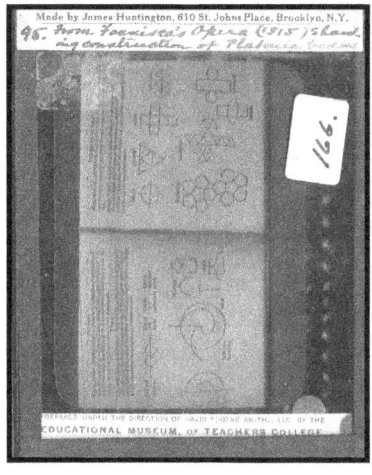

101

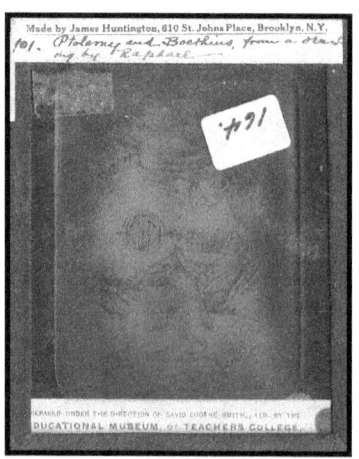

Ptolemy and Boethius, from a drawing by Raphael.

B. Lantern Slides 181

103

Leonardo of Pisa, from an engraving.

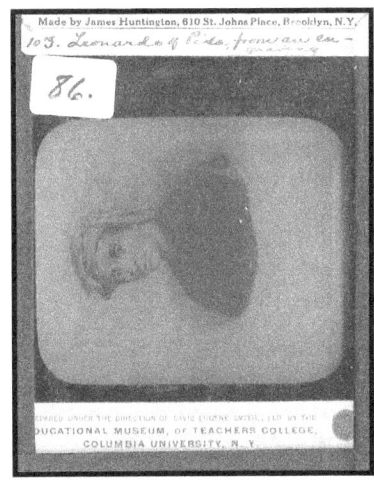

107

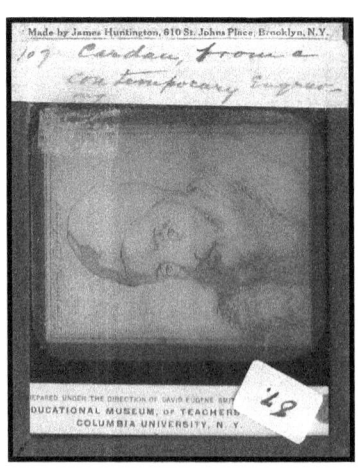

Cardan, from a
contemporary engraving.

B. Lantern Slides 183

108

Tartaglia, from a
contemporary engraving.

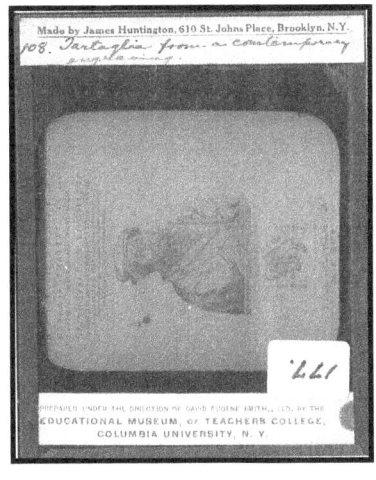

113

Pascal.

B. Lantern Slides

114

Newton.

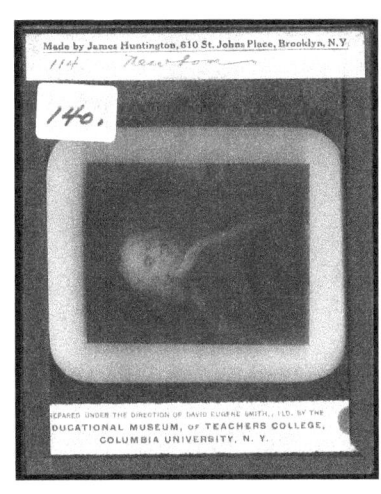

115

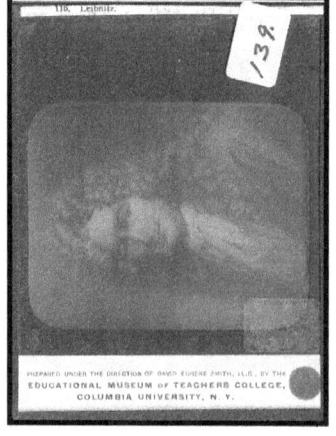

Leibnitz.

B. Lantern Slides 187

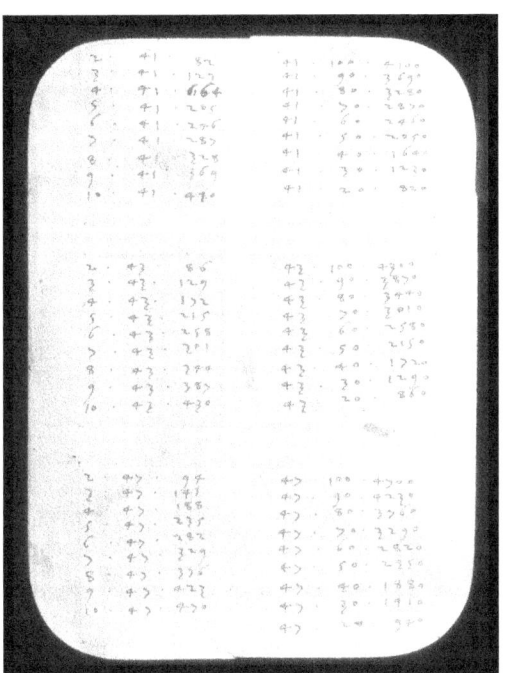

129

Elaborate multiplication table
from a Florentine MS.
(c. 1500).

139

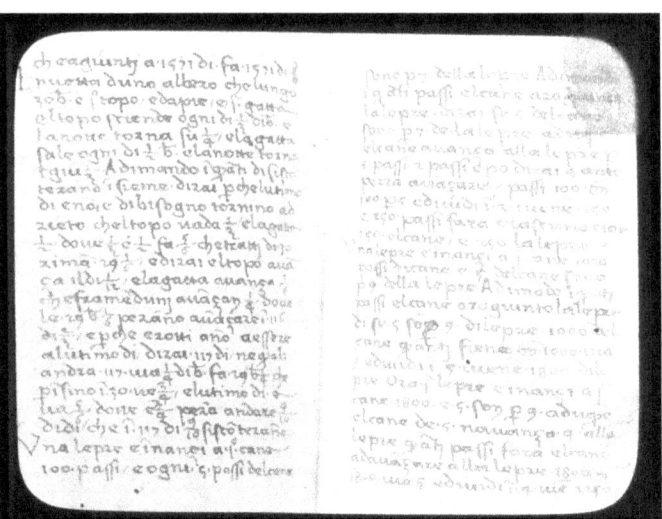

Two pages from a 1460 MS. of the arithmetic of Benedetto di Firenze, showing the hound and hare problem.

B. Lantern Slides 189

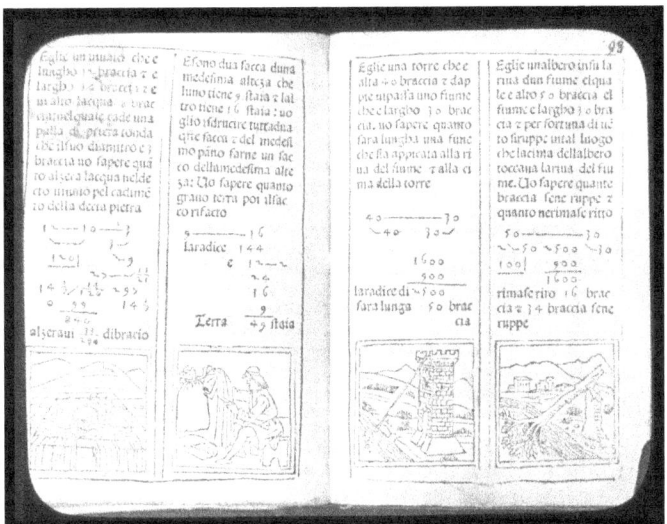

141

From Calandri's arithmetic (1491), showing two pages of illustrated problems. (See No. 70.)

153

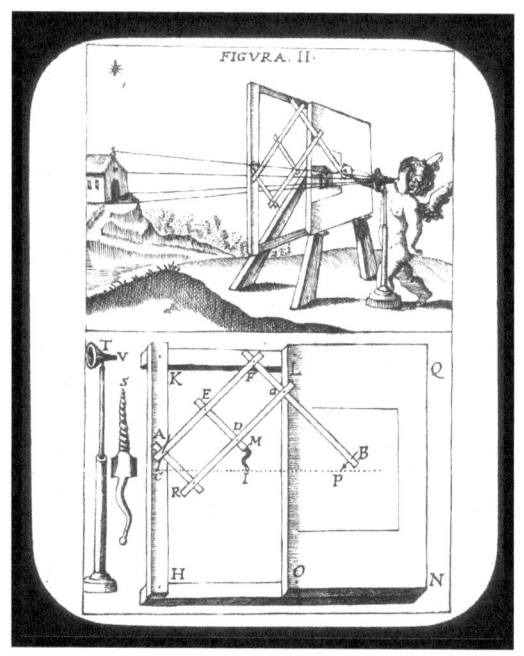

Early use of the *'Parallelogrammo'* (pantograph). From the *Prattica del Parallelogrammo*, by Scheiner (Bologna, 1653).

B. Lantern Slides

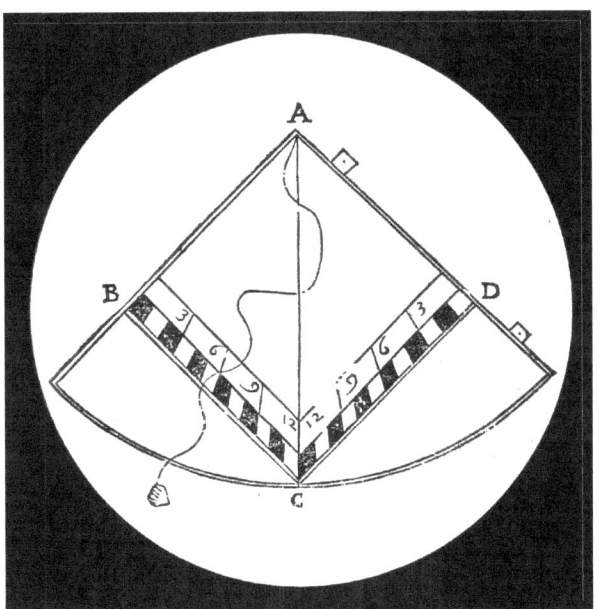

154

From the *Modo di Misurare of Bartoli* (Venice, 1589), showing the uses of the quadrans squadro, baculus, etc.

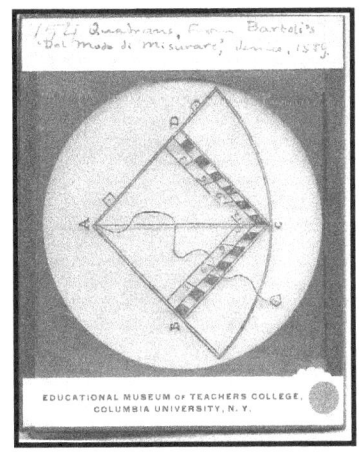

169

Title-page of the first (1494) edition of Paciuolo's *Summa*, showing the topics studied. (See No. 140.)

B. Lantern Slides

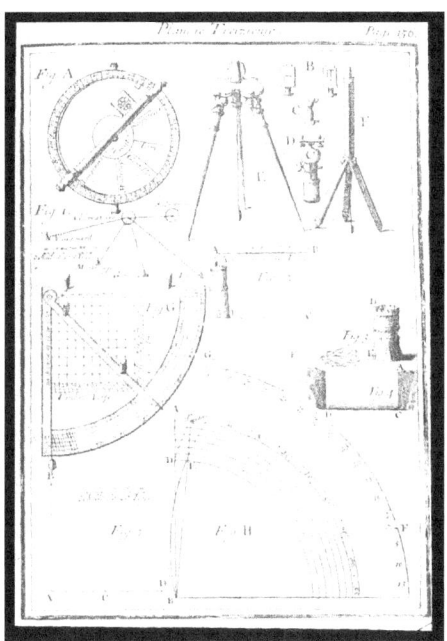

191

From Bion's work on *Instrumens de Mathematique* (La Haye, 1723), showing various mathematical instruments and their uses.

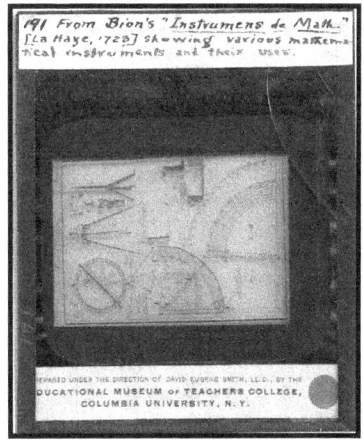

194 B. Lantern Slides

198

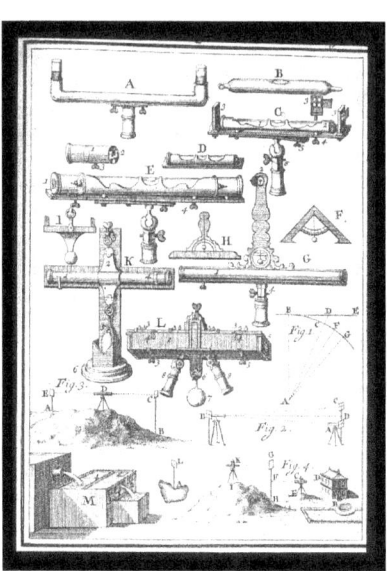

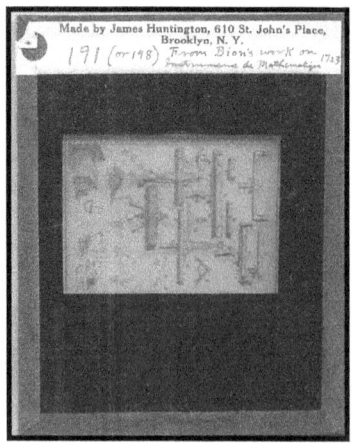

From Bion's work on *Instrumens de Mathematique* (La Haye, 1723), showing various mathematical instruments and their uses.

B. Lantern Slides   195

208 AND 208A

The Comptometer.

209

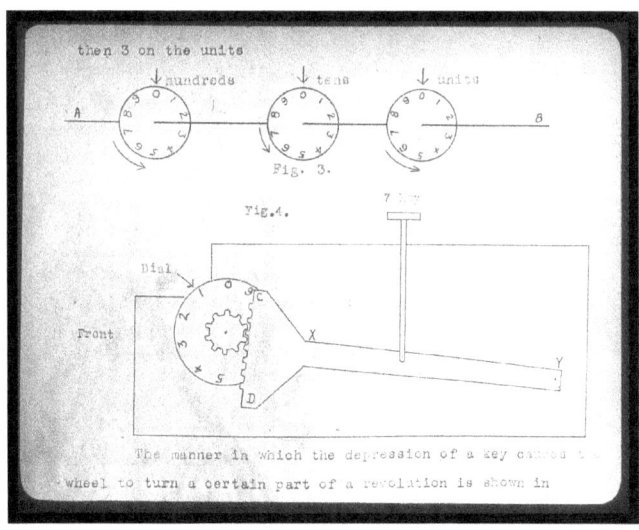

Principle of the Comptometer.

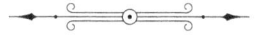

B. Lantern Slides 197

210

Multiplication on the
Comptometer.

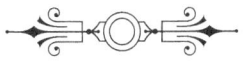

198 B. Lantern Slides

212

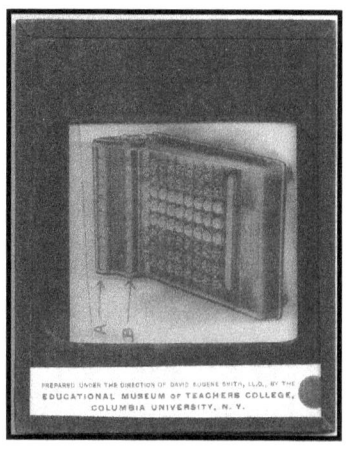

The Mechanical Accountant.

B. Lantern Slides 199

213

Beach Adding Machine.

214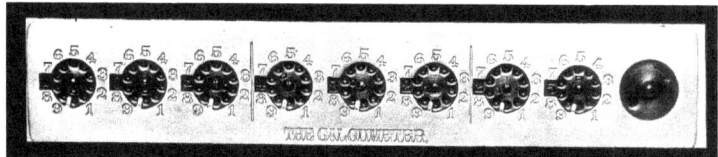

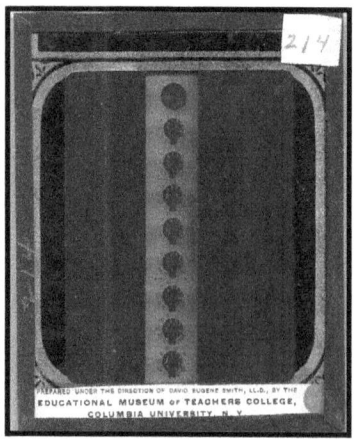

The Calcumeter.

B. Lantern Slides 201

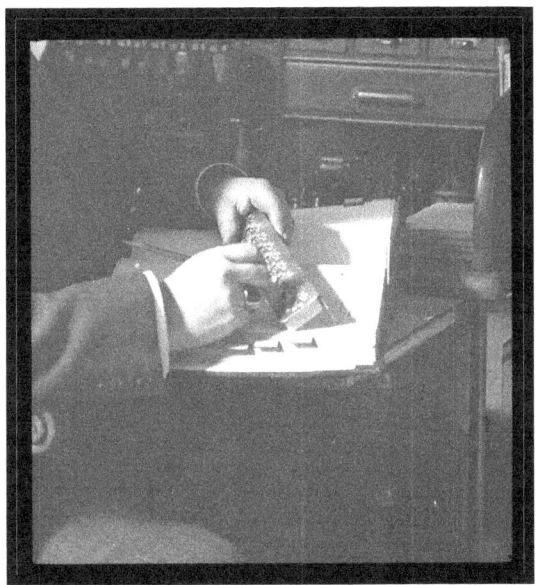

214B

The Calcumeter.

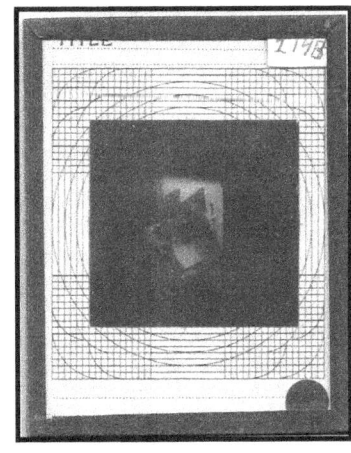

215

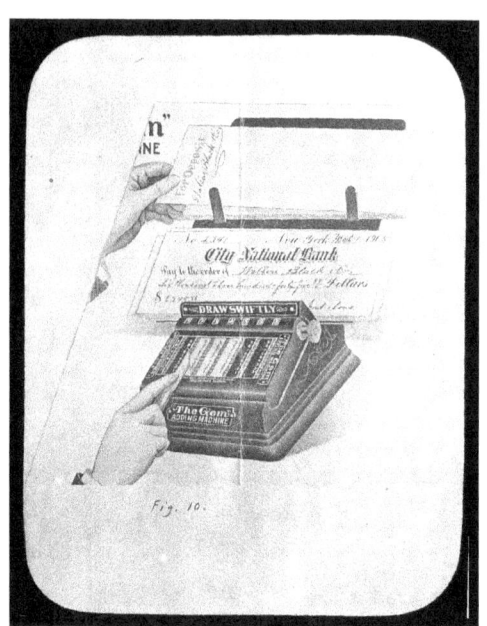

The Gem Adding Machine.

B. Lantern Slides

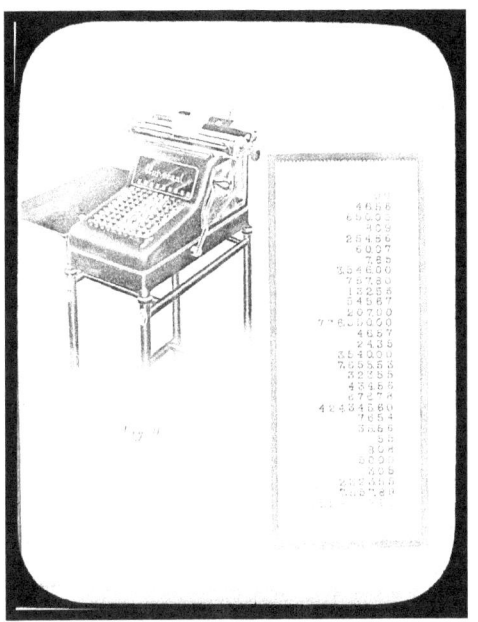

216

The Universal Adding Machine.

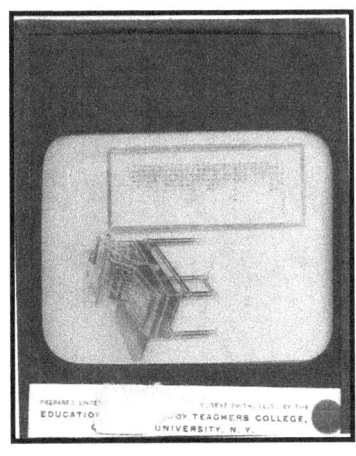

217

Burroughs Adding Machine.

B. Lantern Slides

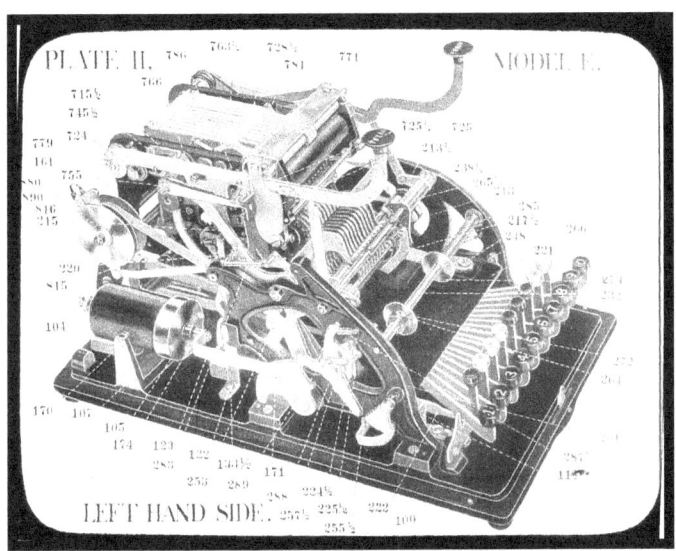

218A

The Standard Adding Machine.

218B

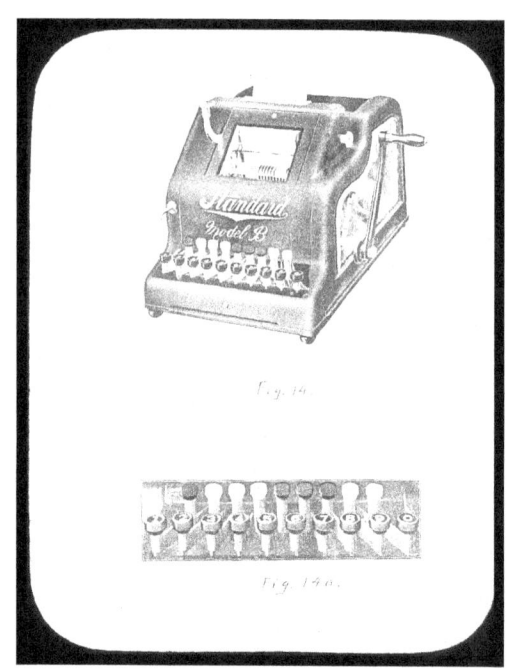

The Standard Adding Machine.

B. Lantern Slides

219

The Comptograph.

220

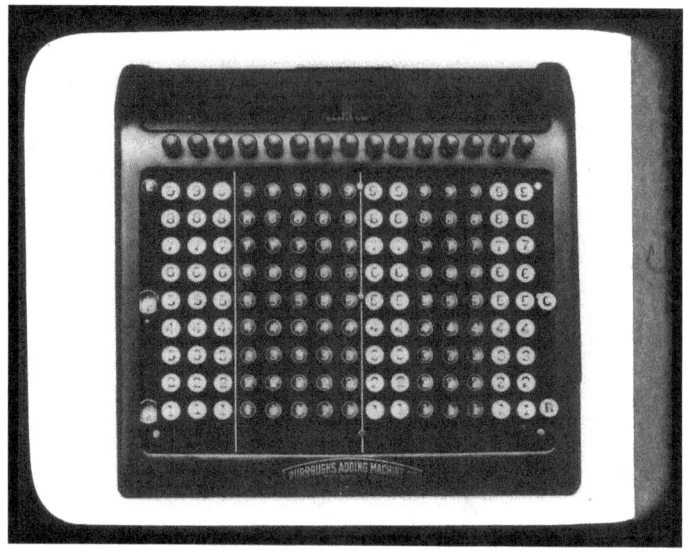

Split key-board of the Burroughs machine.

B. Lantern Slides 209

224

Burroughs Fractional Machine.

B. Lantern Slides

228

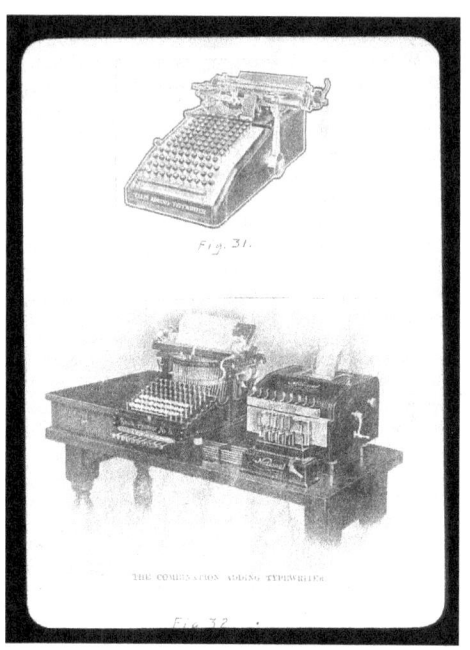

The Ellis Adding Typewriter, and the National Typewriter Adding Machine.

B. Lantern Slides

229

Fig. 33.

The Arithmograph.

B. Lantern Slides

230

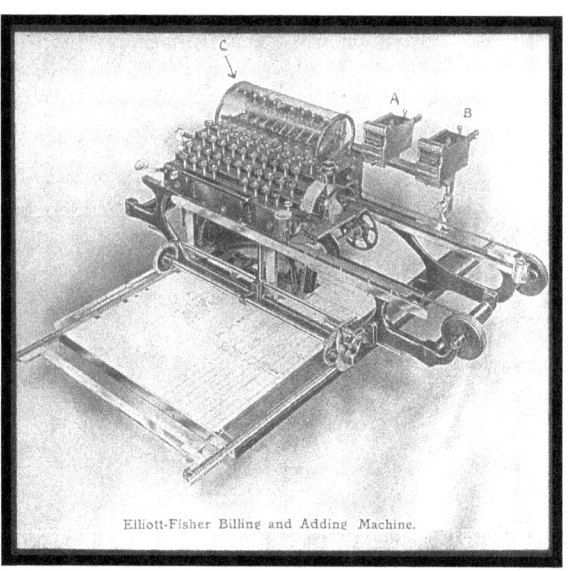

Elliott-Fisher Billing and Adding Machine.

The Elliott-Fisher Billing and Adding Machine.

B. Lantern Slides 213

235

The National Cash Register.

214 B. Lantern Slides

237

The National Cash Register for department stores.

B. Lantern Slides

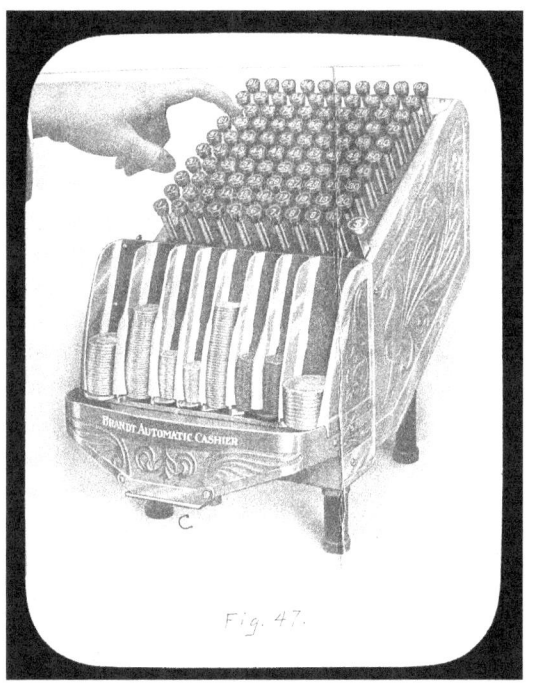

Fig. 47.

239

The Brandt Change-Making Machine.

240

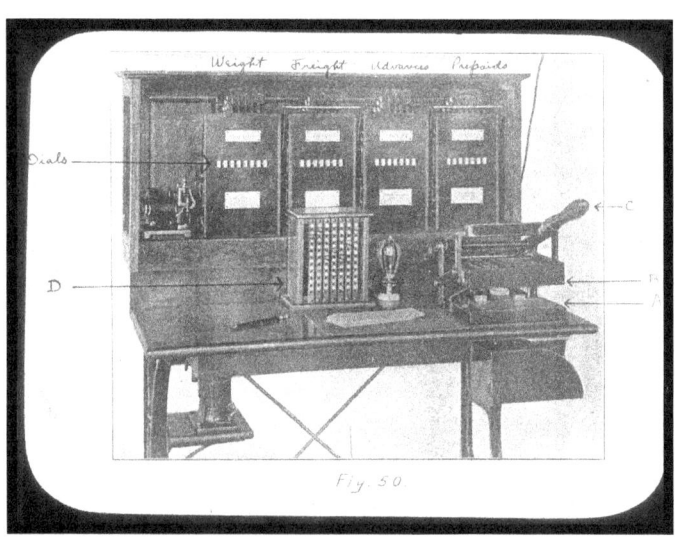

The Hollerith Electric Tabulating and Adding Machine.

B. Lantern Slides 217

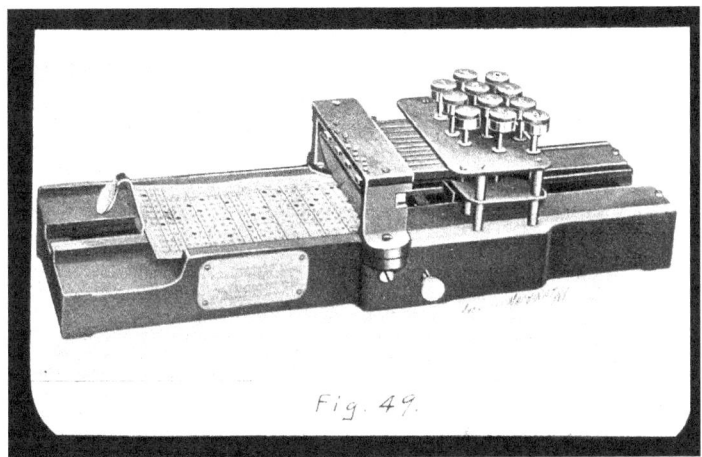

Fig. 49.

242

Punch used with Hollerith machine.

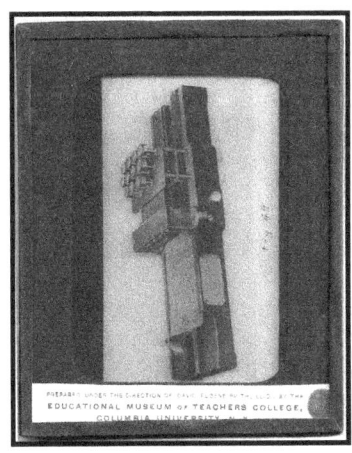

218　　　　　　　B. Lantern Slides

243

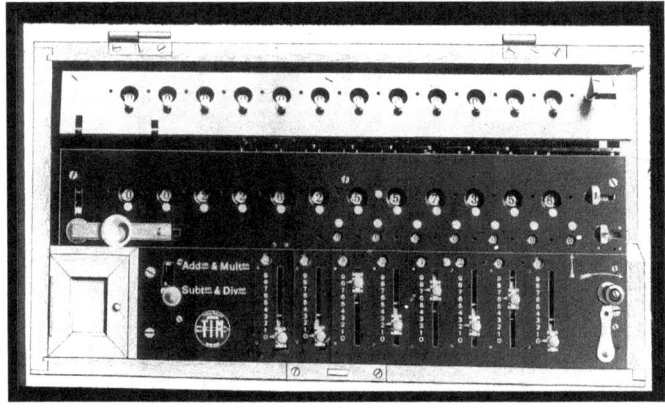

The Saxonia Reckoning Machine.

B. Lantern Slides

245

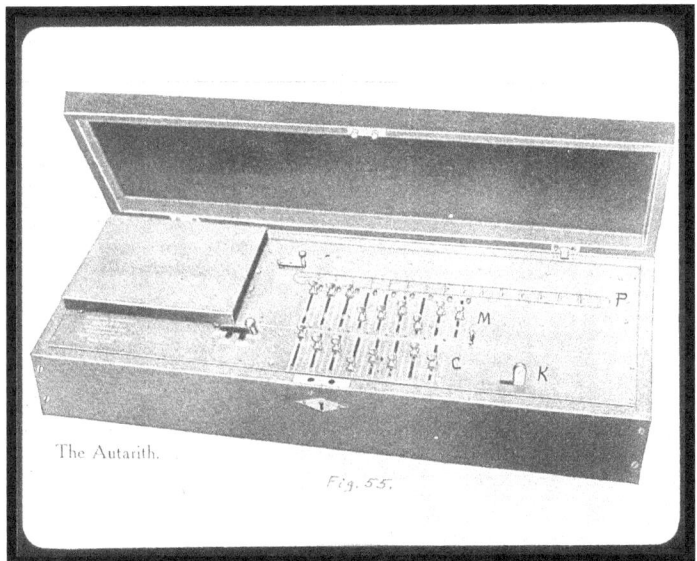

The Autarith.
Fig. 55.

The Autarith.

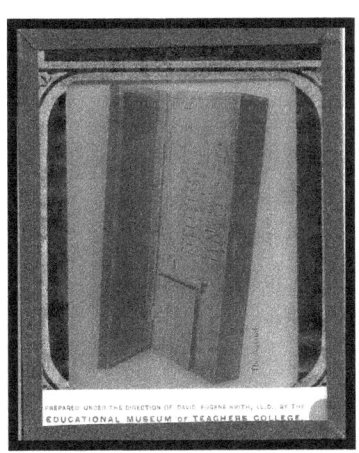

246

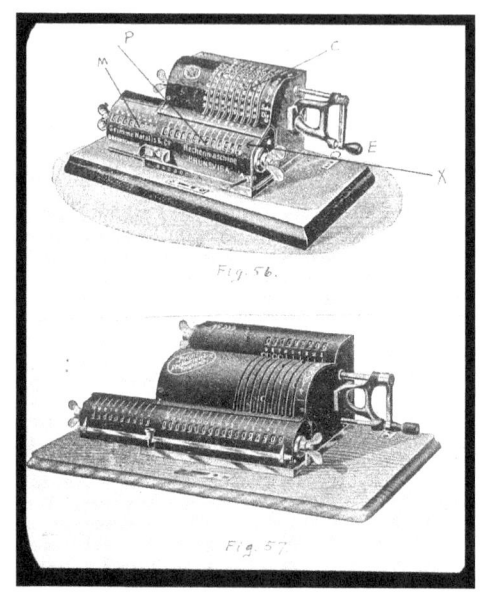

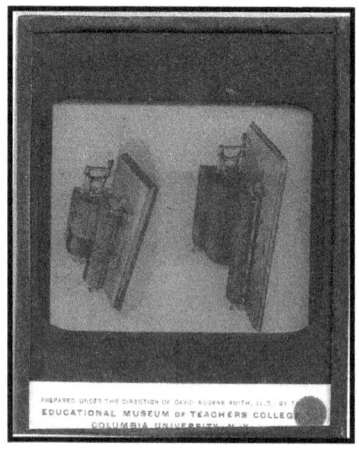

The Brunsviga Calculating Machine. The Triumphator Calculating Machine.

B. Lantern Slides 221

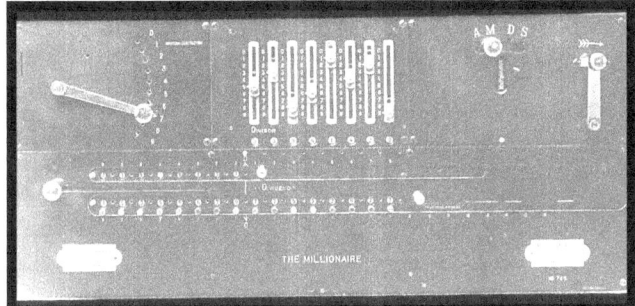

247

The Millionaire Calculating Machine.

247A

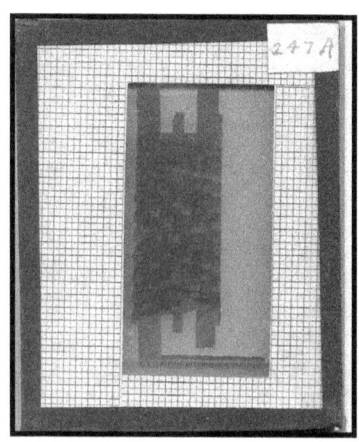

The Millionaire Calculating Machine. [Photo of inner workings.]

●●

B. Lantern Slides

247B

The Millionaire Calculating Machine. [Photo of inner workings.]

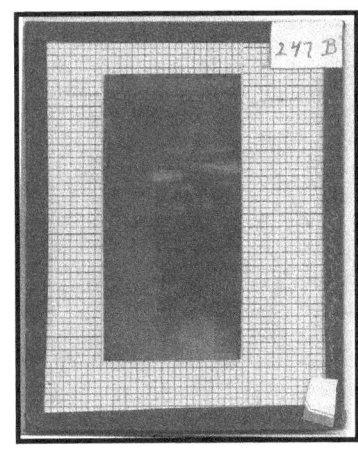

224   B. Lantern Slides

247C

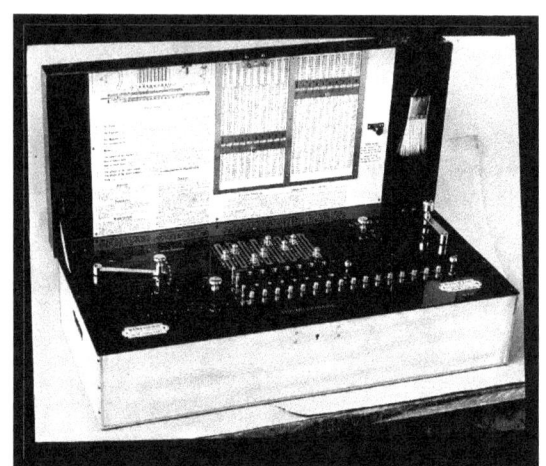

The Millionaire Calculating Machine.

B. Lantern Slides

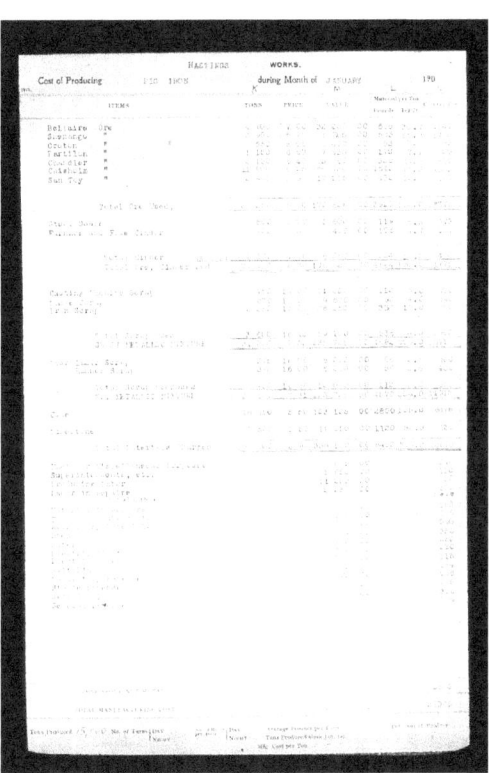

248

Manufacturing cost sheet computed by the multiplying machine.

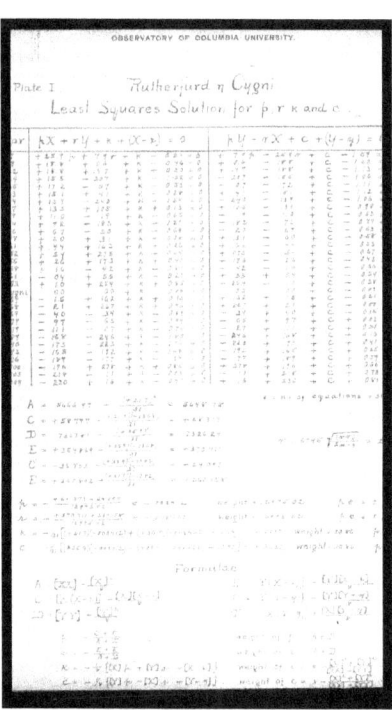

249

Least square solution made by the multiplying machine.

B. Lantern Slides 227

250

The Slide Rule.

B. Lantern Slides

251

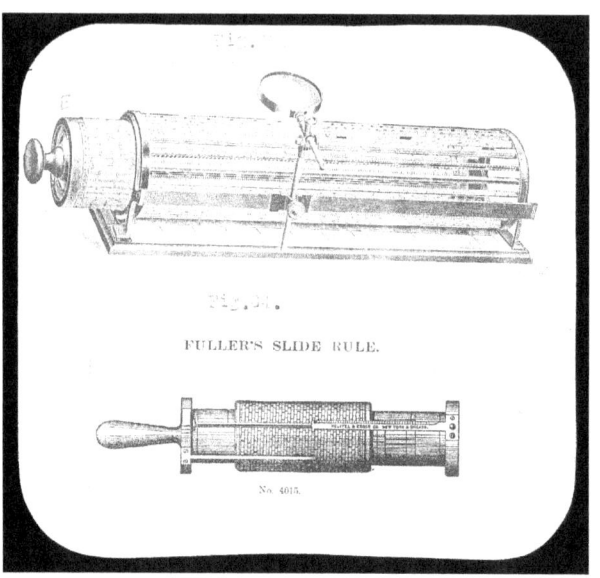

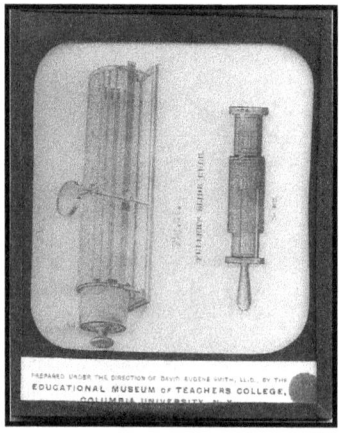

The Thacher Slide Rule. The Fuller Slide Rule.

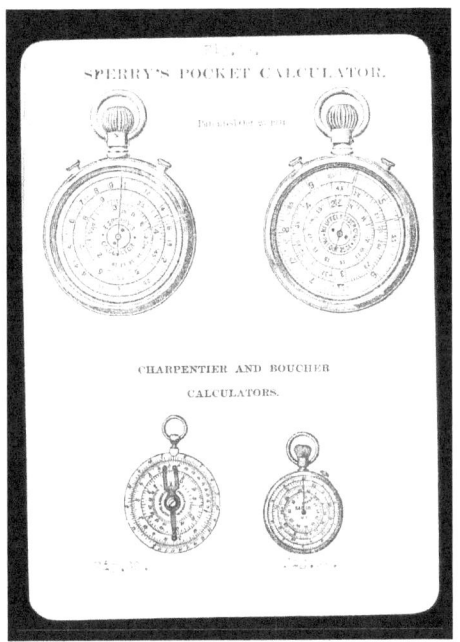

252

Sperry's Pocket Calculator. The Charpentier Calculator. The Boucher Calculator.

253

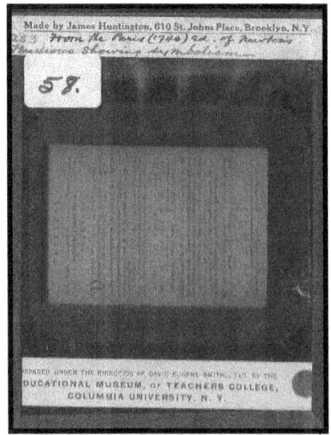

From the Paris (1740) edition of Newton's Fluxions showing symbolism.

B. Lantern Slides

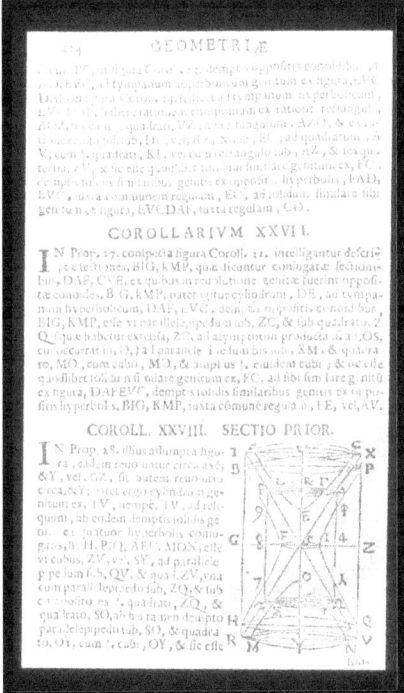

254

From Cavalieri's *Geometria Indivisibilibus* (Bologna, 1653), showing a geometric figure.

256

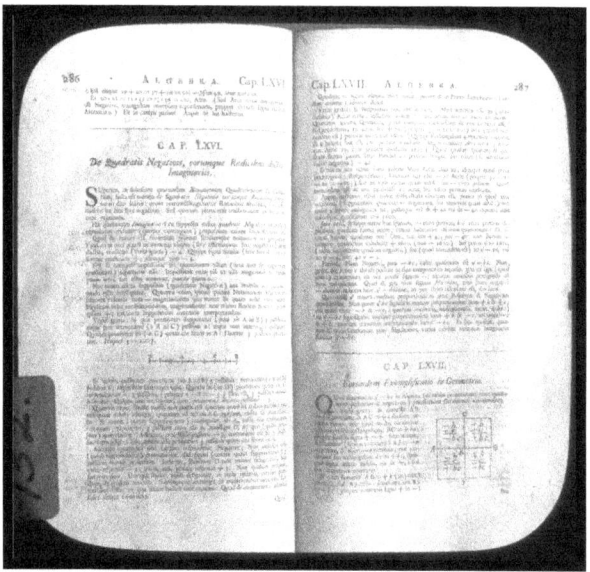

From Wallis's *Tractatus de Algebra* (Oxford, 1685, 1693 edition), showing first geometric treatment of complex number.

B. Lantern Slides

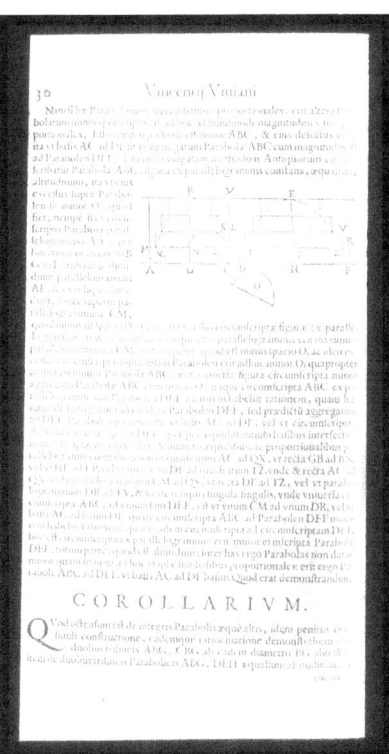

258

From Viviani's *De Max. et Minimis* (Florence, 1659), showing early progress towards calculus.

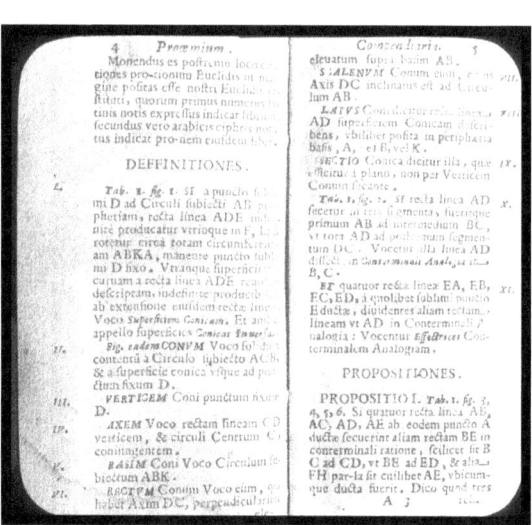

From the *Elementa Conica* of Apollonius (Rome, 1679), showing the nature of the definitions.

B. Lantern Slides 235

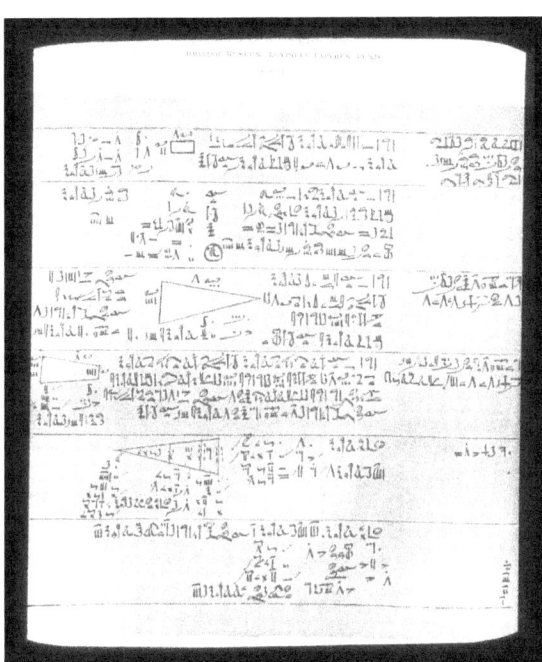

266

From the Ahmes Papyrus.
(See No. 2.)

276

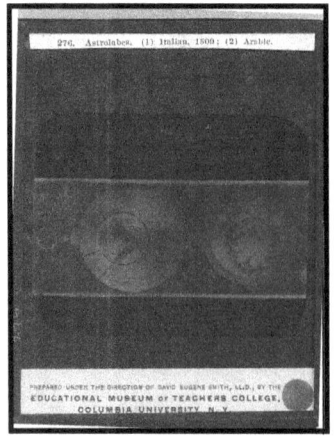

Astrolabes. (1) Italian, 1509; (2) Arabic.

B. Lantern Slides 237

277

Astrolabes. Italian,
1450 and 1558.

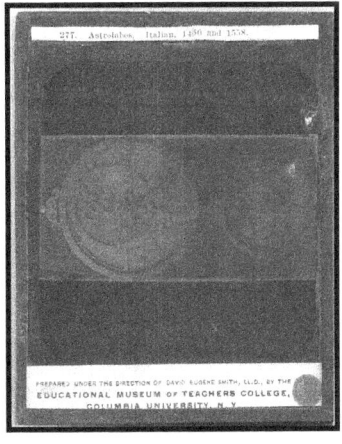

238   B. Lantern Slides

278

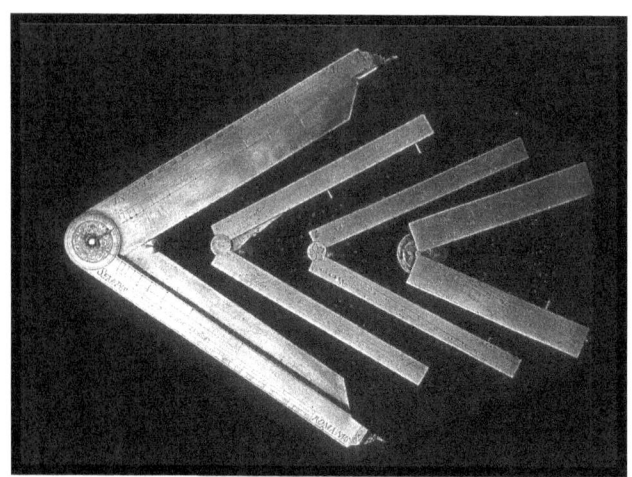

Sector compasses.
Renaissance period.

B. Lantern Slides   239

The remaining slides are not listed in "Illustrations for Lectures on the History of Mathematics." These are most likely taken from items in George A. Plimpton's collection. They are organized using the labels on the slide itself along with the prefix "NL" for not listed.

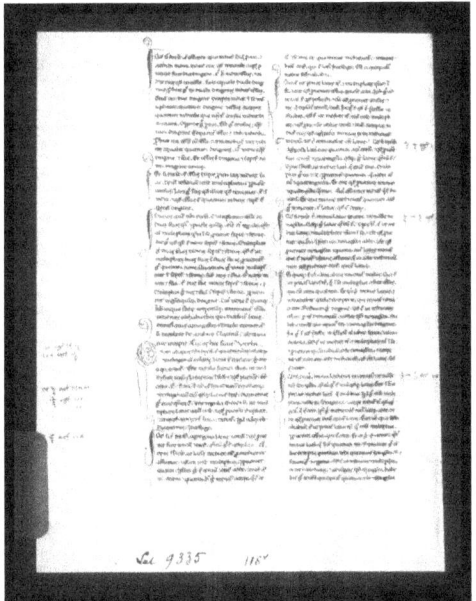

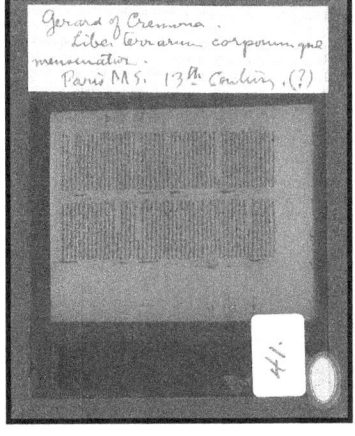

NL 43

BRAHMEGUPTA * CUTTACAD'HYAYA

QUADRATIC EQUATION.

48.¹ RULE for elimination of the middle term:² § 32, 33. Take absolute number from the side opposite to that from which the square and simple unknown are to be subtracted. To the absolute number multiplied by four times the [coefficient of the] square, add the square of the [coefficient of the] middle term; the square root of the same, less the [coefficient of the] middle term, being divided by twice the [coefficient of the] square, is the [value of the] middle term.³

49. Question 16. When does the residue of revolutions of the sun, less one, fall, on a Wednesday, equal to the square root of two less than the residue of revolutions, less one, multiplied by ten and augmented by two?

The value of residue of revolutions is to be here put square of *yávat-távat* with two added: *ya v* 1 *ru* 2 is the residue of revolutions. This less two is *ya v* 1; the square root of which is *ya* 1. Less one, it is *ya* 1 *ru* 1; which multiplied by ten is *ya* 10 *ru* 10; and augmented by two *ya* 10 *ru* 8. It is equal to the residue of revolutions *ya v* 1 *ru* 2 less one: viz. *ya v* 1 *ru* 1. Statement of both sides *ya v* 0 *ya* 10 *ru* 8 Equal subtraction being made
ya v 1 *ya* 0 *ru* 1

NL 44

NL 45

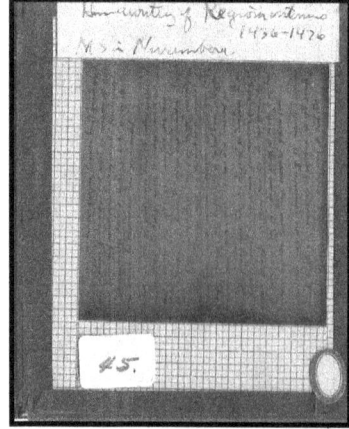

B. Lantern Slides

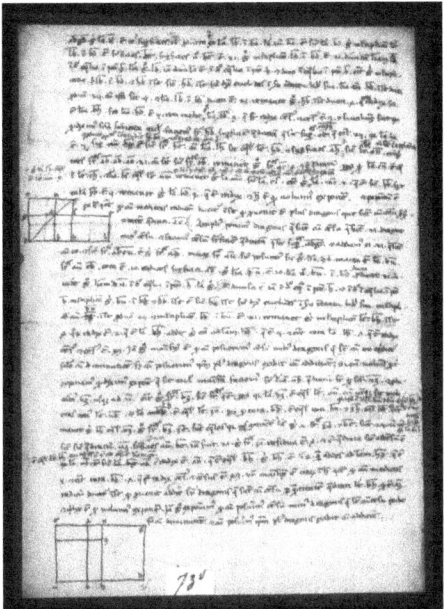

NL 46

NL 47

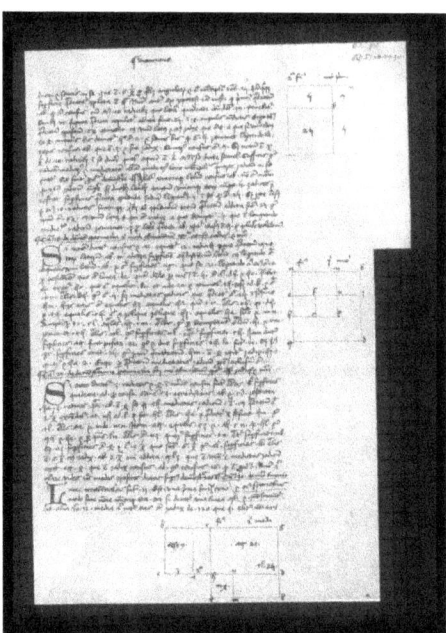

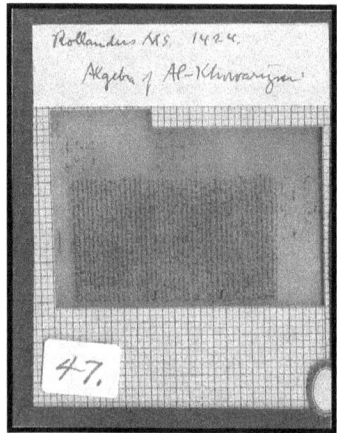

B. Lantern Slides

NL 48

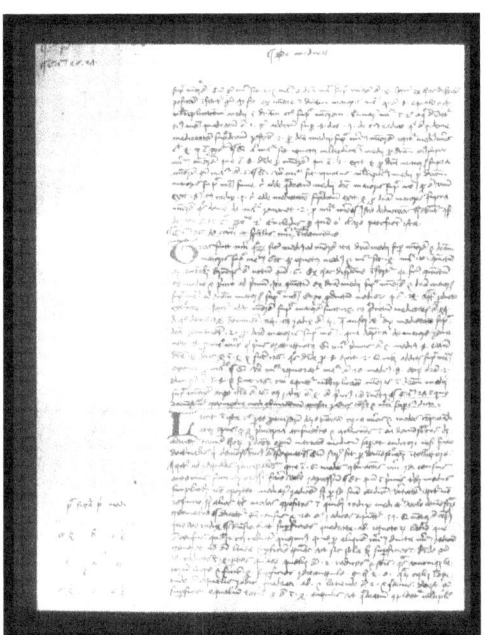

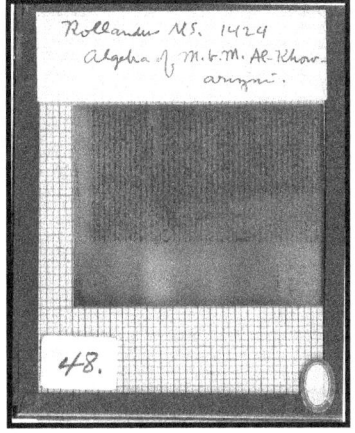

NL 49

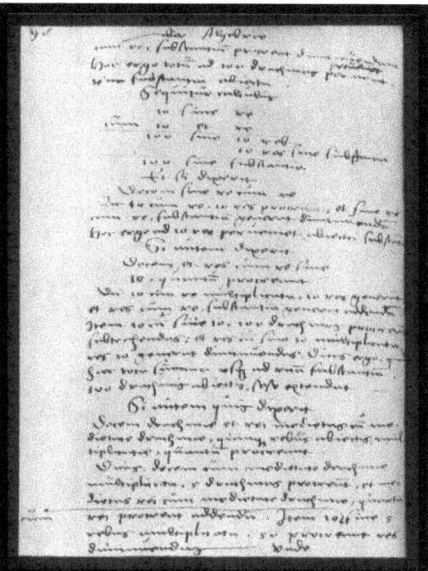

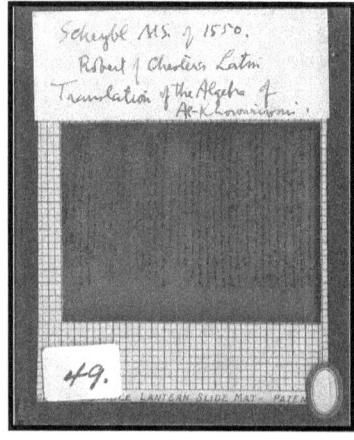

B. Lantern Slides 247

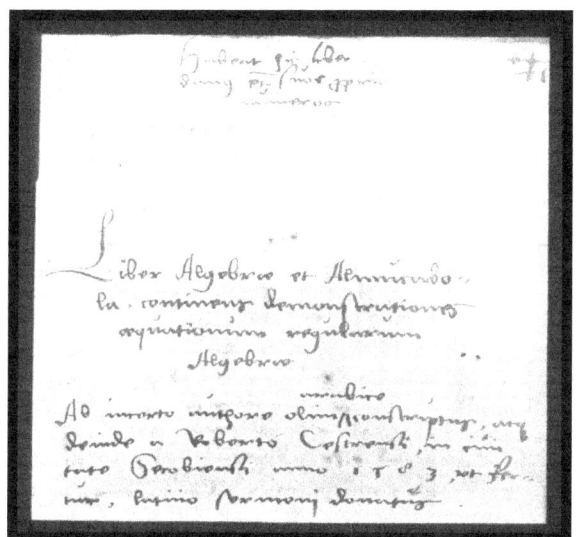

NL 50

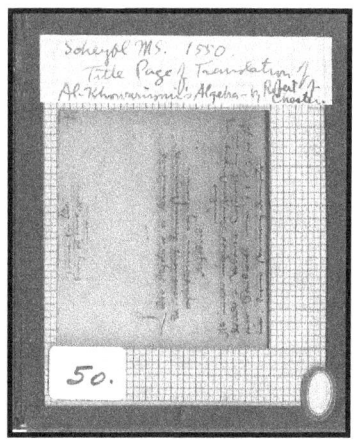

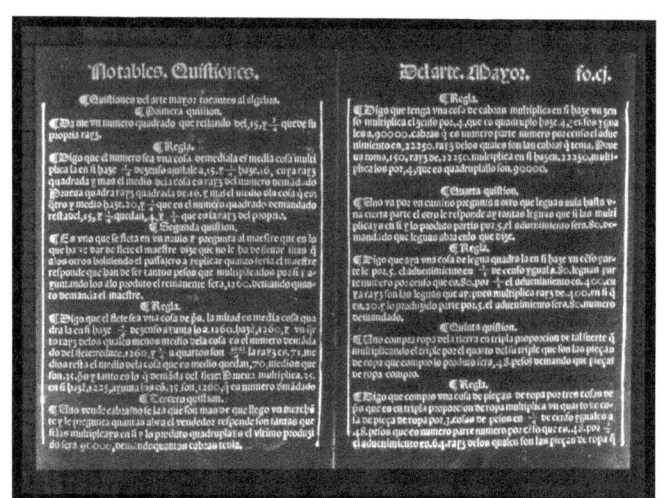

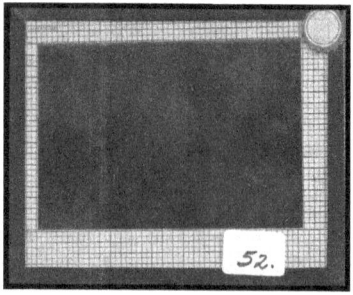

NL 53

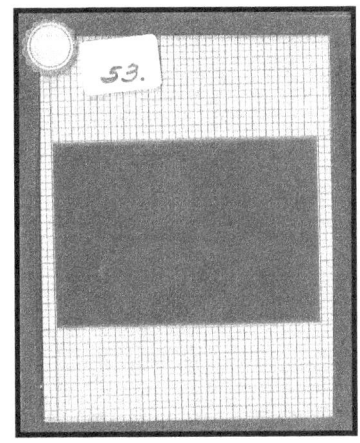

NL 54

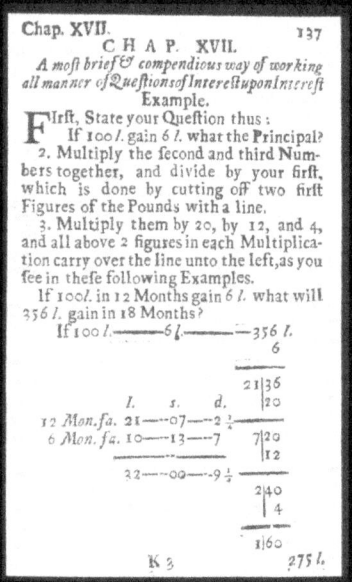

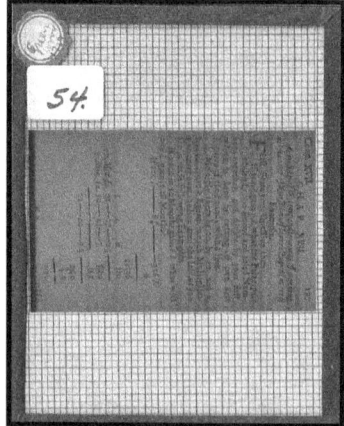

B. Lantern Slides

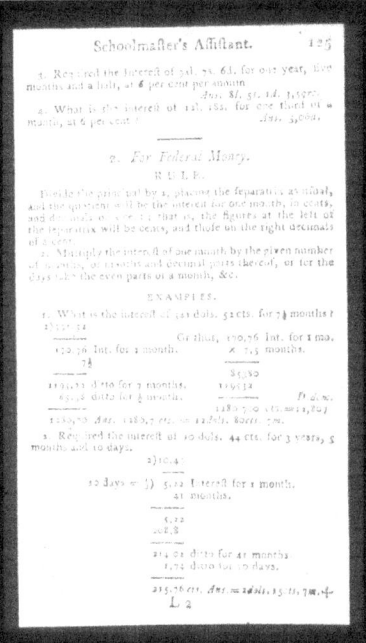

NL 56

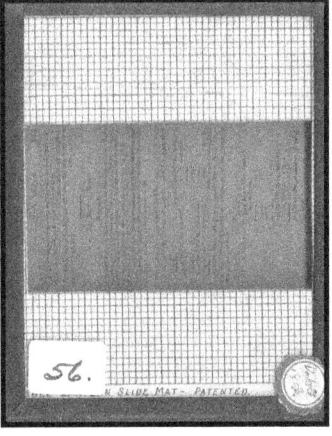

NL 57

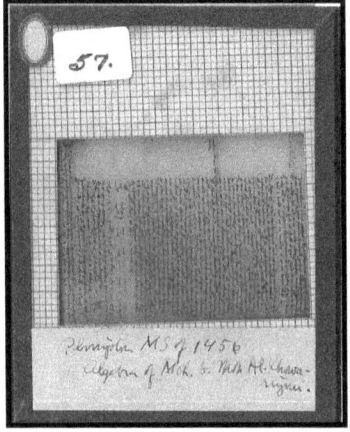

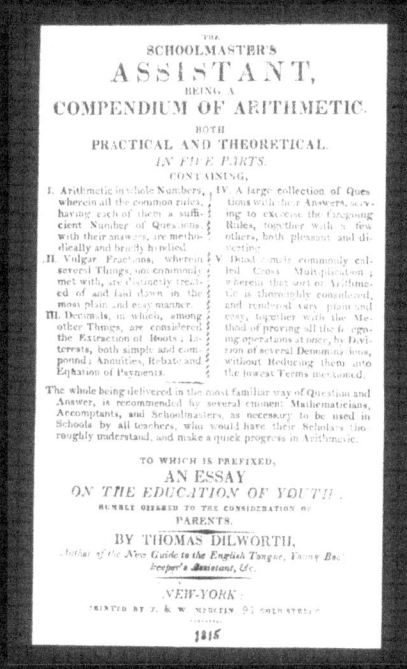

NL 60

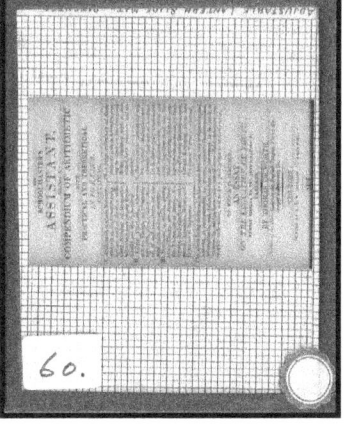

NL 63

CONTENTS.

	Page
II. BARTER.	123
Two Examples.	123, 124
III. EQUATION of PAYMENTS.	124 to 126
The *Common Rule*.	124
A *Necessary Observation*.	126
IV. LOSS, and GAIN.	126
Four Cases.	126, 127
V. ALLEGATION.	128
Medial.	128, 129
Alternate: Three Cases.	129 to 133
Allegation Partial.	131
Allegation Total.	132
VI. POSITION, or the RULE of FALSE.	134
Single.	134
Double.	134 to 137
VII. INTEREST.	137
Simple; Four Propositions.	138 to 140
Compound; Four Propositions.	140 to 142
VIII. REBATE, or DISCOUNT.	143
Simple.	143
Compound.	143, 144
IX. ANNUITIES.	144
1. *Annuities, at Simple Interest*.	144 to 149
Annuities in Arrears; Four Propositions.	144 to 147
The Present Worth of Annuities; Four Propositions.	147 to 149
2. *Annuities, at Compound Interest*.	149 to 154
Annuities in Arrears; Four Propositions.	149 to 152
The Present Worth of Annuities; Four Propositions.	152 to 154
X. Of REVERSIONS, and FREEHOLD ESTATES.	154 to 158
Annuities in Reversion; Two Propositions.	154 to 156
The Purchase of Freehold Estates, Five Propositions.	156 to 158
Dr. *Halley's Table* of the *Value* of *Annuities* upon *Life*.	158

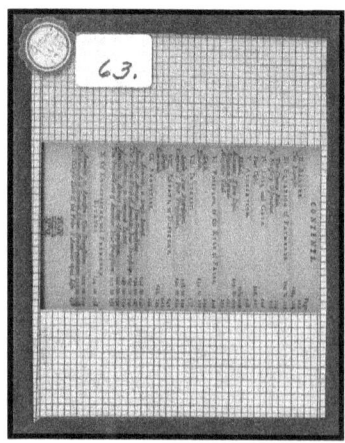

B. Lantern Slides

NL 64

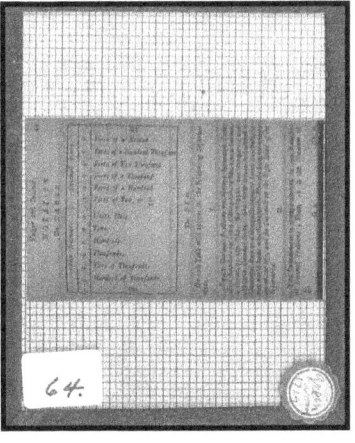

NL 68

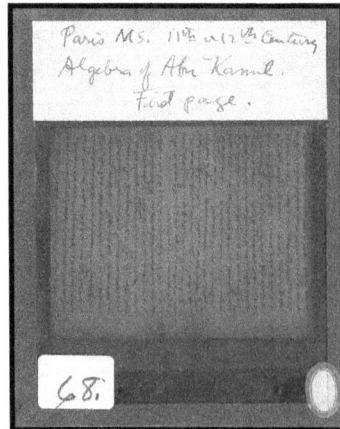

Paris MS. 11th or 12th Century
Algebra of Abu Kamil.
First page.

68.

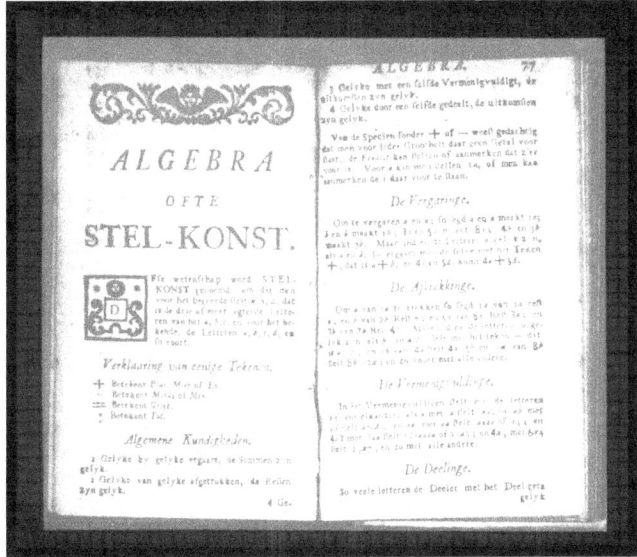

NL 73

NL 75

Chap. V.   *Division.*   55

I *shall not,* I *(hope) need to trouble my self, or* Learner, *to shew the Working of this Sum, or any other, having, now (as* I *suppose) sufficiently treated of Division; but will leave it to the Censure of the experienc'd to judge, whether this Manner of dividing be not plain, lineal,* & *to be wrought with fewer Figures than any which is commonly taught :* As for Example appeareth.

```
           (8
          97 (5
         9863 (0
        987529 (3
       9876418x (0
      98765x2609 (8
     987654x95987 (6
    4938x7x4848765 (4
   x469x35786376543 (2
  xx345678998765432x (124999999
  98765432xxxxxxxxx   98765432x
  98765432xxxxxxxx    124999999
  9876543333333       2499999982
  98765444444         3749999974
  987655555           4999999966
  9876666             6249999958
  98777               7499999940
  988                 8749999933
  8                   9999999920
                      11249999915
                              8
Proof  123456789987654321
E 2                        CHAP.
```

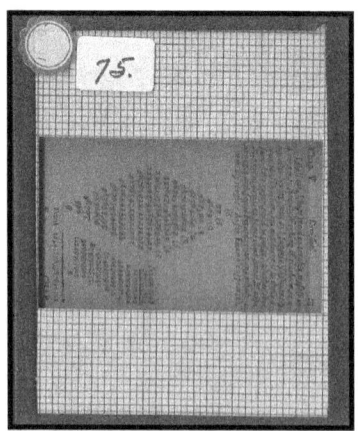

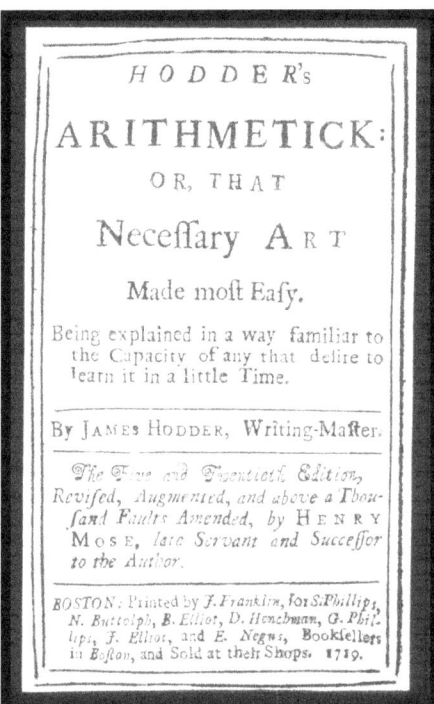

NL 76

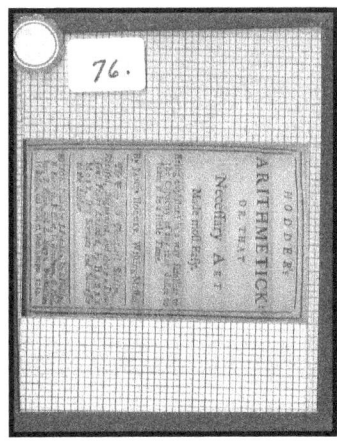

NL 77

THE
Young Mathematician's Guide.
Being a PLAIN and EASIE
INTRODUCTION
TO THE
Mathematicks.
In Five PARTS.
Viz.

I. **Arithmetick**, Vulgar and Decimal, in all its Useful Rules; With a general Method of Extracting the Roots of all Single Powers.

II. **Algebra**, or Arithmetick in Species; wherein the Method of Raising and Resolving Æquations is rendered Easie; and Illustrated with Variety of Examples, and Numerical Questions. Also the Whole Business of Interest and Annuities, &c. fully and plainly Handled, with several New Improvements.

III. The **Elements** of **Geometry**, Contracted, and Analytically Demonstrated; With a New and Easie Method of finding the Circle's Periphery and Area to any assigned Exactness, by one Æquation only; Also a New Way of making Sines and Tangents.

IV. **Conick-Sections**, wherein the Chief Properties &c. of the Ellipsis, Parabola, and Hyperbola, are Clearly Demonstrated.

V. The **Arithmetick** of **Infinites** Explain'd, and render'd Easie; with its Application to Superficial, and Solid Geometry.

With an
APPENDIX of **Practical Gauging**.

By JOHN WARD. Teacher of the **Mathematicks**.
Heretofore Chief Surveyor and Gauger General in the Excise.

LONDON:
Printed by *Edw. Midwinter*, for *John Taylor* at the *Ship* in St. *Paul's Church-Yard.* 1707.
Price Bound 6 s.

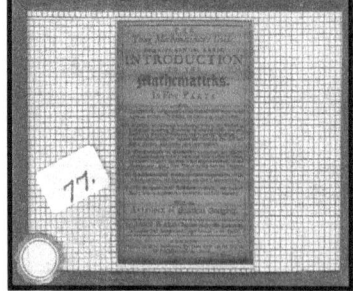

B. Lantern Slides

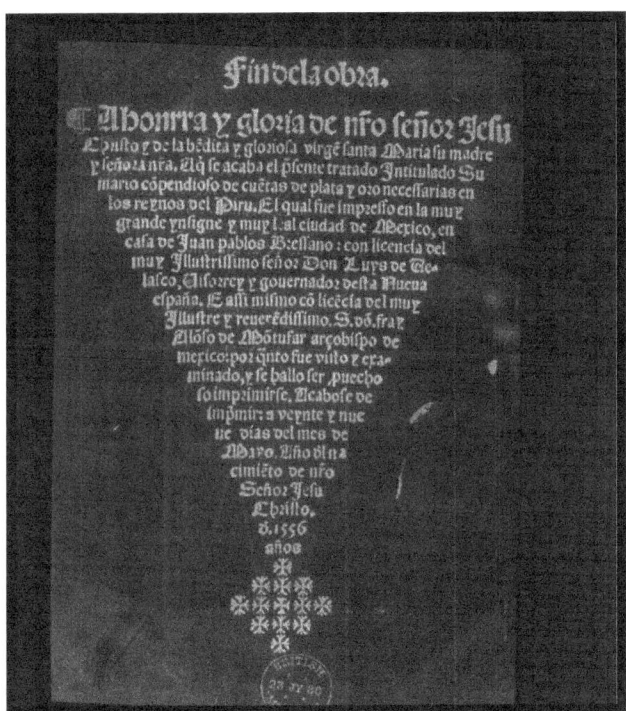

NL 78

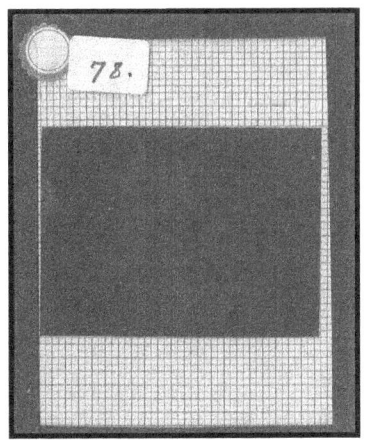

NL 79

ARITHMETICK
Vulgar and *Decimal* ;

WITH THE

APPLICATION
THEREOF, TO
A VARIETY of CASES
IN
Trade, and *Commerce*.

BOSTON: N.E.
Printed by S. KNEELAND and T. GREEN, for T.
HANCOCK at the Sign of the Bible and Three
Crowns in *Annstreet*. MDCCXXIX.

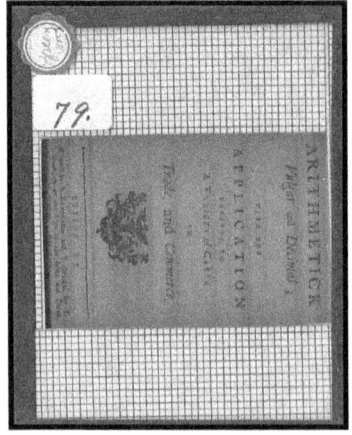

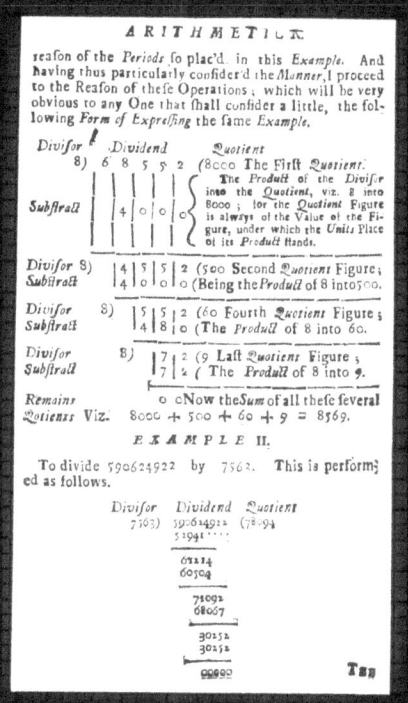

NL 80

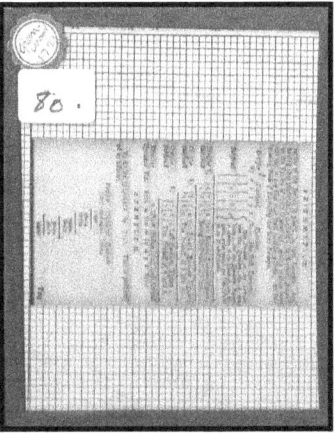

NL 81

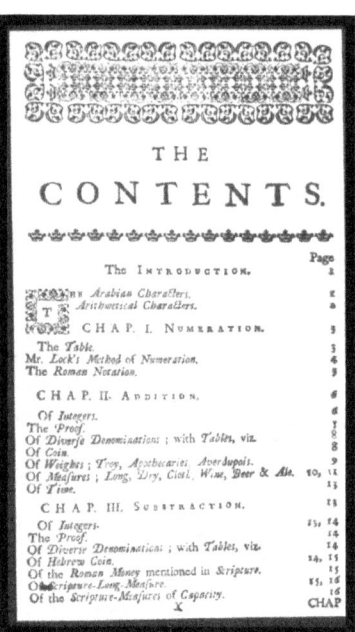

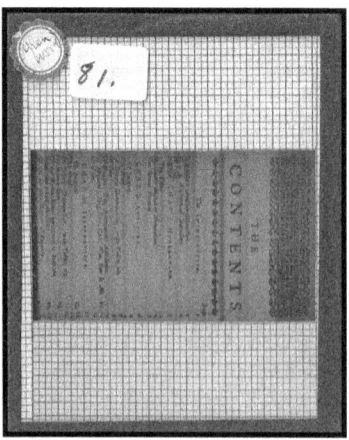

B. Lantern Slides 265

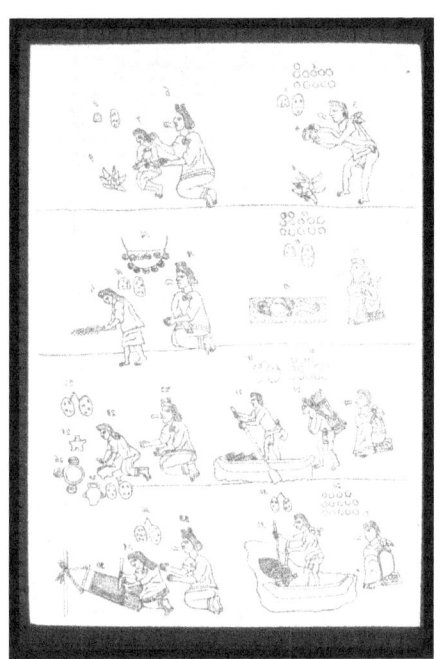

NL 82

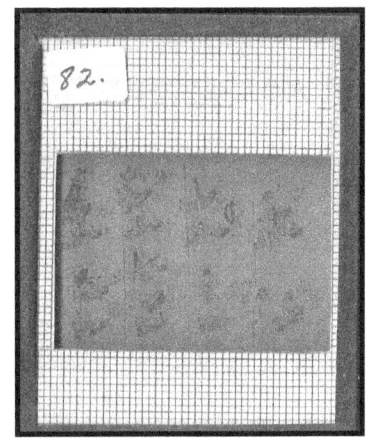

NL 83

Tare and Tret.

Tare is an allowance made to the buyer for the weight of the hogshead, barrel, box, or whatever else contains the goods bought, and is calculated at so much per hogshead, barrel, &c. or at so much per cent, or at so much in the grofs weight.

Tret is an allowance made to the buyer of 4 pounds in 104 for waste and duft in some sorts of goods.

112 pounds weight is call'd a grofs hundred, and 100 pounds a neat hundred; some forts of goods are fold by one weight and fome by the other. When an article is fold by grofs hundreds, the price is generally specified at fo much per hundred, and the tare per cent. is upon 112 pounds. When an article is fold by neat hundreds, the price is generally specified at fo much per pound, and the tare per cent. is upon 100 pounds.

The whole weight of an article, and the hogshead or whatever contains it, being weighed together, is called the grofs weight, whether the article be fold by grofs hundreds or neat hundreds.

The weight of the article itself, after all allowances are deducted, is called the neat weight, whether the article be fold by grofs hundreds or neat.

Cafe 1ft. When the tare is at fo much per hogshead, barrel, &c. multiply the number of hogsheads or barrels by the tare, and the product will be neat hundreds; reduce this product to grofs hundreds if the article is specify'd in grofs hundreds, and subtract it from the grofs weight; the remainder is the neat weight.

Cafe 2d. When the tare is at fo much per cent. and is the aliquot part or parts of an hundred weight, divide the whole grofs by the said part or parts which the tare is of an hundred weight; the quotient thence arifing

gives

[14]

Rebate or Discount.

Rebate or discount is when a sum of money due at any time to come, is satisfy'd by paying so much present money, as being put out to interest, would amount to the given sum in the same space of time.

Find the amount of £100 for the time and rate per cent. given, which interest add to £100; then by a stating in the rule of three say, as that sum is to £100 so is the debt or sum proposed to the present worth required. The difference between the present worth and the given sum is the rebate.

Equation of Payments.

When several sums of money are to be paid at different times, and it is required at what time the whole shall be paid together, without loss to debtor or creditor; this is called equation of payments, or equating the time of payment. Multiply each payment by its time, add the products together, and divide this sum by the whole debt, the quotient is the equated time.

Fellowship.

By Fellowship the accompts of several partners, trading in a company are so adjusted or made up, that every partner may have his just part of the gain, or sustain his just part of the loss; according to the proportion or share of money he hath in the joint stock. There are two kinds of fellowship, viz. single and double. Single fellowship is when the stocks of all the partners continue an equal term of time, and is usually call'd fellowship without time. Double fellowship is when the stocks continue an unequal term of time, and

NL 84

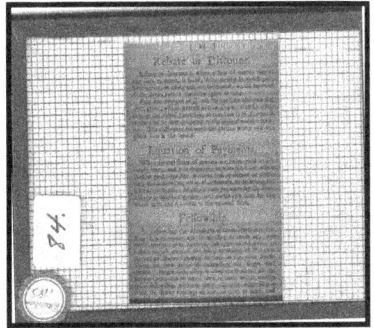

NL 85

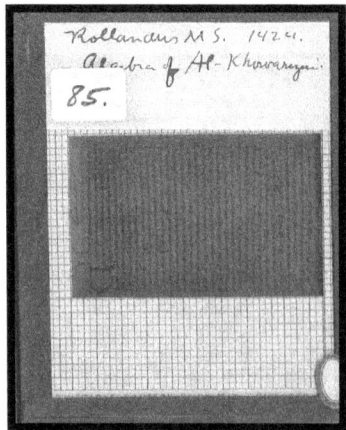

NL 93

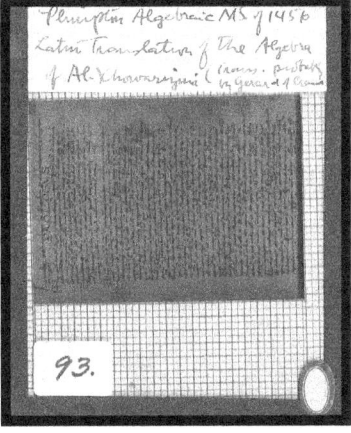

NL 94

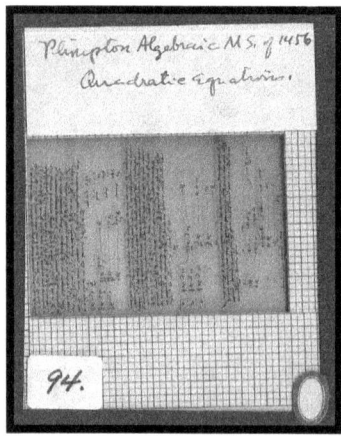

NL 95

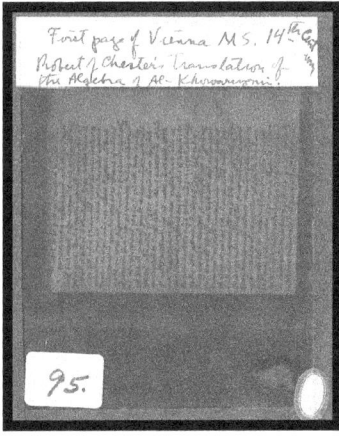

NL 99

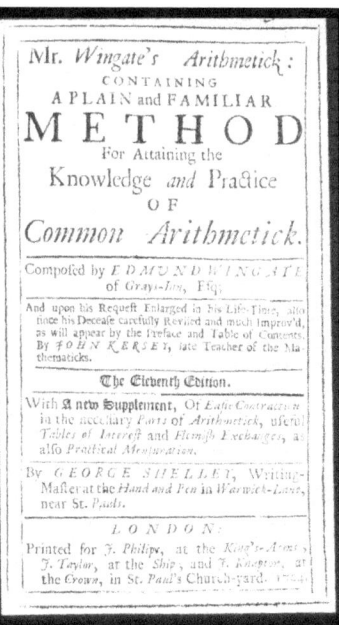

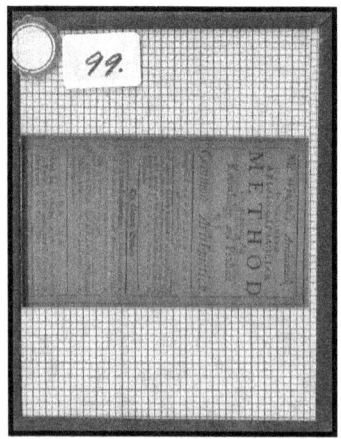

B. Lantern Slides

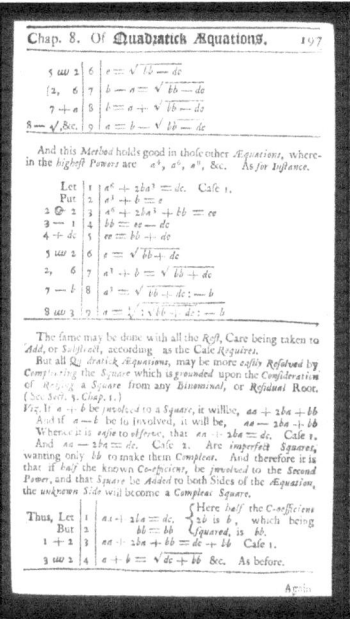

NL 100

NL 126

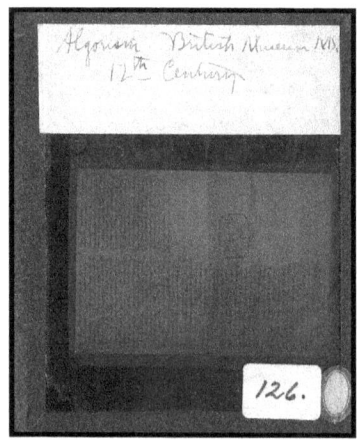

The rule relating to the *Mūla* variety (of miscellaneous problems on fractions):—

33. Half of (the coefficient of) the square root (of the unknown quantity) and (then) the known remainder should be (each) divided by *one* as diminished by the fractional (coefficient of the unknown) quantity. The square root of (the (sum of) the) known remainder (so treated), as combined with the square (of the coefficient) of the square root (of the unknown quantity dealt with as above), and (then) associated with (the similarly treated coefficient of) the square root (of the unknown quantity), and (thereafter) squared (as a whole), gives rise to the (required unknown) quantity in this *mūla* variety (of miscellaneous problems on fractions).

Examples in illustration thereof.

34. One-fourth of a herd of camels was seen in the forest; twice the square root (of that herd) had gone on to mountain-slopes; and 3 times 5 camels (were), however, (found) to remain on the bank of a river. What is the (numerical) measure of that herd of camels?

NL 128

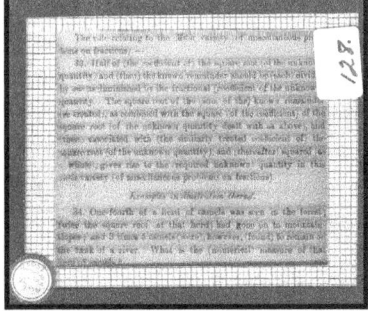

NL 131

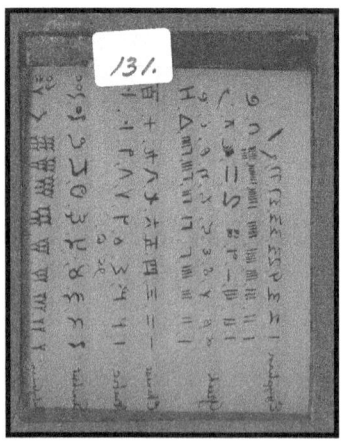

B. Lantern Slides

NL 132

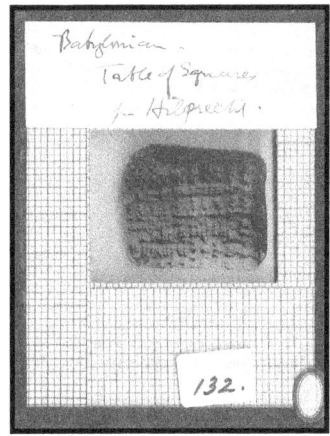

NL 133

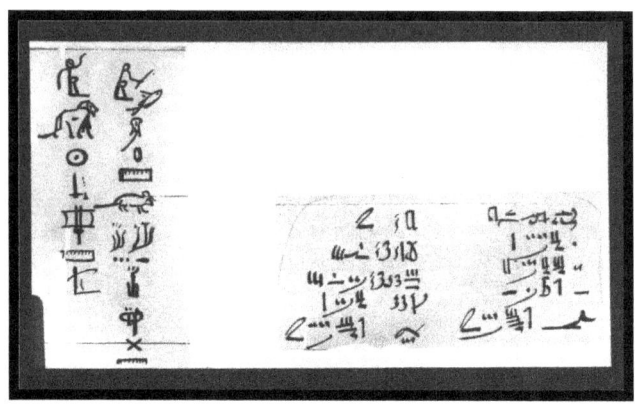

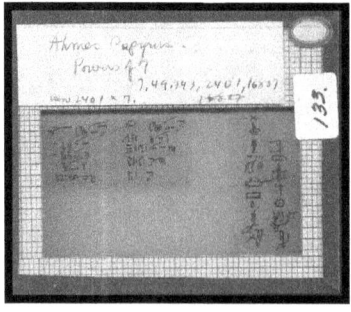

B. Lantern Slides

NL 134

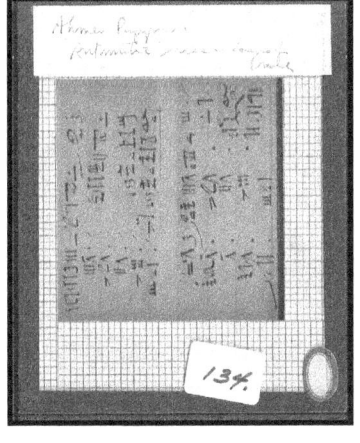

NL 135

$$a : x :: x : y :: y : b,$$ whence $x^2 = ay,$ $y^2 = bx,$ and $xy = ab.$

ARCHIMEDES: ON THE SPHERE AND THE CYLINDER.
II.4. TO CUT A GIVEN SPHERE BY A PLANE SO THAT THE VOLUMES OF THE SEGMENTS ARE TO ONE ANOTHER IN A GIVEN RATIO.

$$x^2(a - x) = b^2 c,$$ whence $x^3 - ax^2 + b^2 c = 0.$

Solved by intersection of the conics, $x^2 = \dfrac{b^2}{a} y,$ and $(a - x)y = ac.$

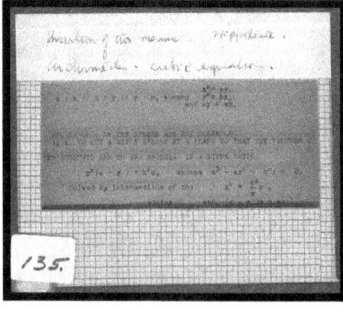

B. Lantern Slides

> **PROPOSITION 29.**
>
> *To a given straight line to apply a parallelogram equal to a given rectilineal figure and exceeding by a parallelogrammic figure similar to a given one.*
>
> Let AB be the given straight line, C the given rectilineal figure to which the figure to be applied to AB is required to be equal, and D that to which the excess is required to be similar;
>
> thus it is required to apply to the straight line AB a parallelogram equal to the rectilineal figure C and exceeding by a parallelogrammic figure similar to D.
>
>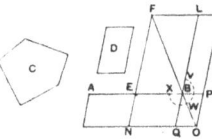
>
> Let AB be bisected at E;
>
> let there be described on EB the parallelogram BF similar and similarly situated to D;
>
> and let GH be constructed at once equal to the sum of BF, C and similar and similarly situated to D. [vi. 25]
>
> Let KH correspond to FL and KG to FE.
> Now, since GH is greater than FB,
>
> therefore KH is also greater than FL, and KG than FE.

NL 141

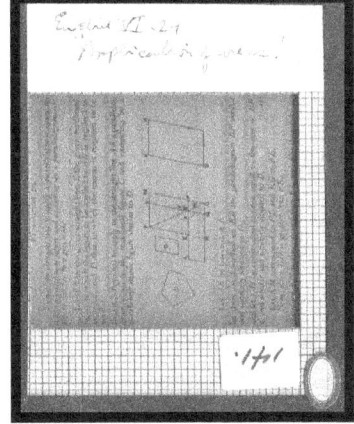

PROPOSITION 30.

To cut a given finite straight line in extreme and mean ratio.

Let AB be the given finite straight line;

thus it is required to cut AB in extreme and mean ratio.

On AB let the square BC be described;

and let there be applied to AC the parallelogram CD equal to BC and exceeding by the figure AD similar to BC. [VI. 29]

Now BC is a square;

therefore AD is also a square.

And, since BC is equal to CD,

let CE be subtracted from each;

therefore the remainder BF is equal to the remainder AD.

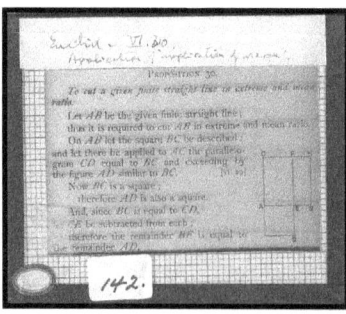

We cannot have a real solution of this unless
$$x > \tfrac{1}{8}(x^2-60) \text{ and } < \tfrac{1}{5}(x^2-60).$$
Therefore $\quad 5x < x^2-60 < 8x.$
(1) Since $\quad x^2 > 5x+60,$
$x^2 = 5x +$ a number greater than 60,
whence x is[1] *not less than* 11.
(2) $\quad x^2 < 8x+60$
or $\quad x^2 = 8x +$ some number less than 60,
whence x is[1] *not greater than* 12.
Therefore $\quad 11 < x < 12.$
Now (from above) $x = (m^2+60)/2m$;
therefore $22m < m^2+60 < 24m.$
Thus (1) $\quad 22m = m^2 +$ (some number less than 60),
and therefore m is[2] *not less than* 19.
(2) $24m = m^2 +$ (some number greater than 60),
and therefore m is[2] *less than* 21.
Hence we put $m = 20,$ and
$$x^2-60 = (x-20)^2,$$
so that $x = 11\tfrac{1}{2},\ x^2 = 132\tfrac{1}{4},$ and $x^2-60 = 72\tfrac{1}{4}.$
Thus we have to divide $72\tfrac{1}{4}$ into two parts such that $\tfrac{1}{8}$ of one part *plus* $\tfrac{1}{5}$ of the other $= 11\tfrac{1}{2}.$
Let the first part be $5z.$
Therefore $\tfrac{1}{8}$ (second part) $= 11\tfrac{1}{2}-z,$
or second part $= 92-8z$;
therefore $5z+92-8z = 72\tfrac{1}{4}.$
and $z = \dfrac{79}{12}.$

NL 148

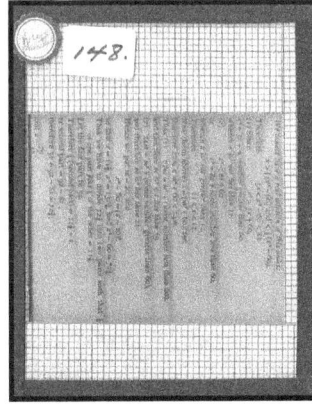

NL 149

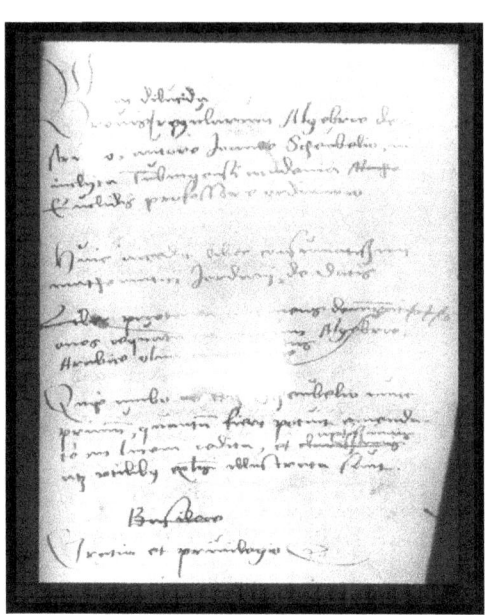

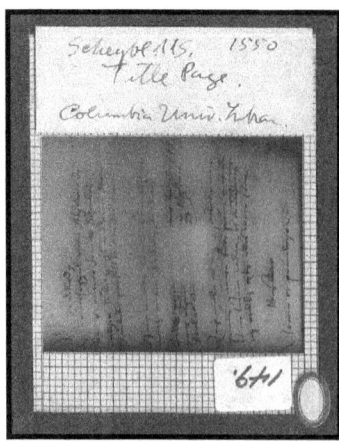

B. Lantern Slides

NL 150

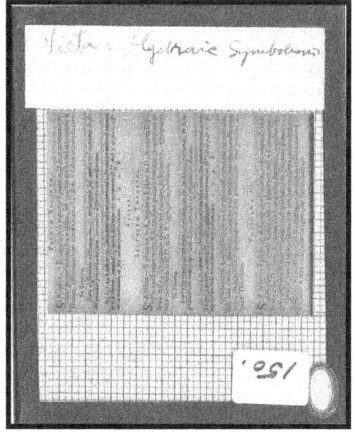

NL 156

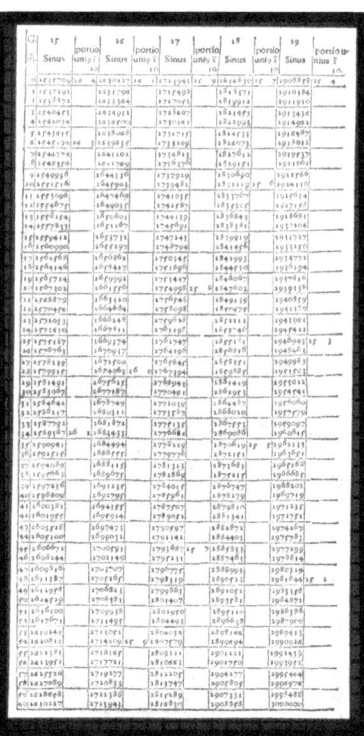

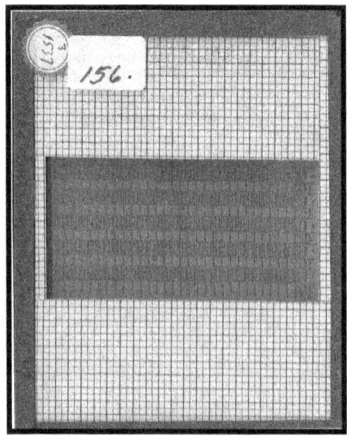

B. Lantern Slides

Sinttsrecti.

g\m	24	25	26	27	28		29	
	ptes	ptes	ptes	ptes	ptes		ptes	
31	24897	15846	26787	27720	28644		29560	
32	24913	62	16803	35	60		75	
33	29	77	19	51	75		90	
34	45	93	262	34	27766	90	26906	
35	24960	25909	16840	82	28706	255	21	
36	76	25	65	97	21		63	
37	92	40	81	27813	36		29651	
38	25008	164	25956	96	28	28752	66	
39	24	71	26912	44	67		81	
40	40	88	27	27859	82		97	
41	25056	26003	43	75	98		29712	
42	72	19	26959	90	28813		17	
43	87	35	74	27905	28		42	
44	25103	26051	90	21	44		29757	
45	19	66	27005	36	257	28859	72	
46	35	82	21	27952	74		88	
47	25151	98	37	67	89		29803	252
48	67	26113	17052	83	28905		19	
49	82	29	63	98	20		33	
50	98	45	83	28014	35		29843	
51	25214	26161	99	29	51		63	
52	30	76	27114	44	28966		78	
53	46	92	30	28060	81		94	
54	25162	26208	46	259	96		29909	
55	77	23	17061	91	29012		24	
56	93	39	77	28106	27		39	
57	25309	26255	92	22	42		29954	
58	25	70	27203	37	29058		69	
59	41	86	23	52	73		84	
60	57	26302	39	68	88		30000	

NL 157

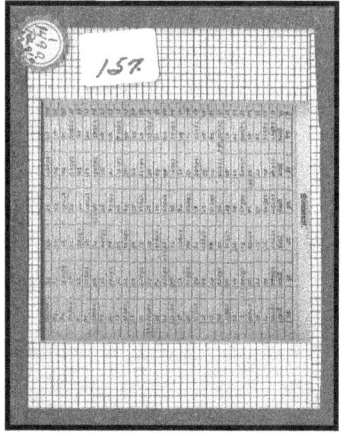

NL 158

	Tabula Secunda				
	Numerus		Numerus		Numerus
B		B		B	
0	00000	31	60086	61	180402
1	1745	32	62486	62	188075
2	3492	33	64940	63	196263
3	5240	34	67452	64	205034
4	6992	35	70022	65	214450
5	8748	36	72654	66	224607
6	10511	37	75356	67	235583
7	12278	38	78129	68	247513
8	14053	39	80978	69	260511
9	15838	40	83909	70	274753
10	17633	41	86929	71	290422
11	19439	42	90040	72	307767
12	21256	43	93254	73	327088
13	23087	44	96571	74	348748
14	24932	45	100000	75	373211
15	26794	46	103551	76	401089
16	28674	47	107236	77	433148
17	30573	48	211062	78	470453
18	32492	49	115037	79	514438
19	34433	50	119177	80	567118
20	36396	51	123491	81	631377
21	38387	52	127994	82	711569
22	40402	53	132704	83	814456
23	42448	54	137639	84	951387
24	44522	55	142813	85	1143131
25	46631	56	148253	86	1430203
26	48772	57	153987	87	1908217
27	50952	58	160015	88	2863563
28	53170	59	166410	89	5724706
29	55432	60	173207	90	Infinitu
30	57734				

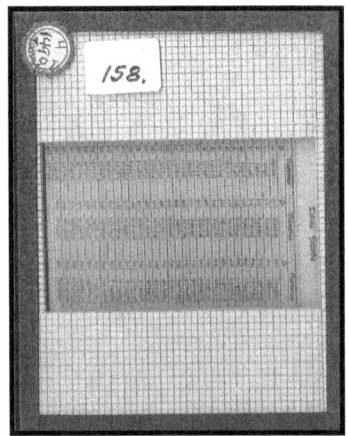

B. Lantern Slides

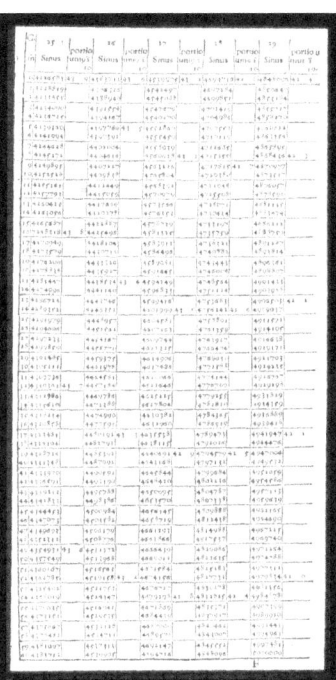

NL 159

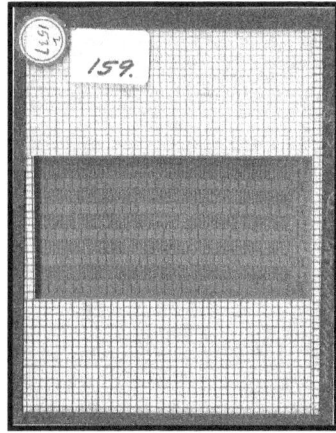

NL 163

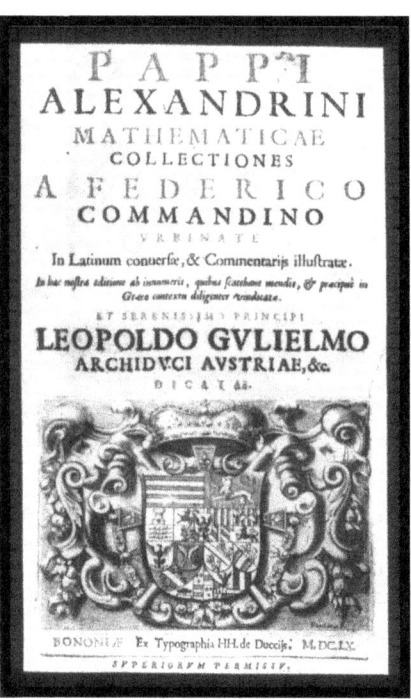

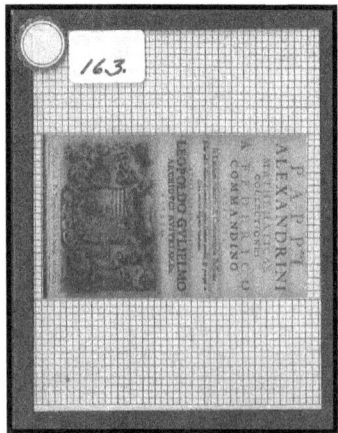

B. Lantern Slides 291

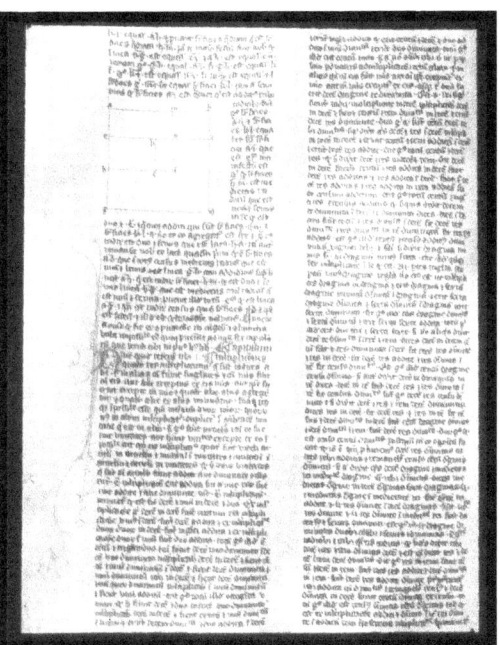

NL 169

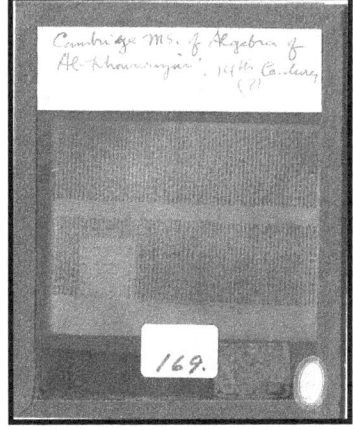

NL 171

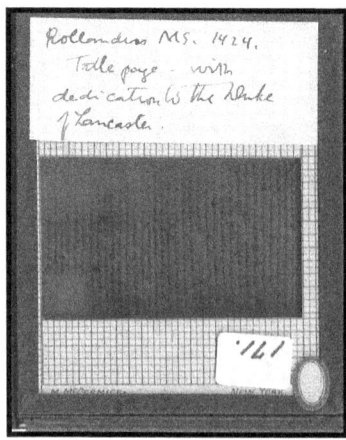

B. Lantern Slides

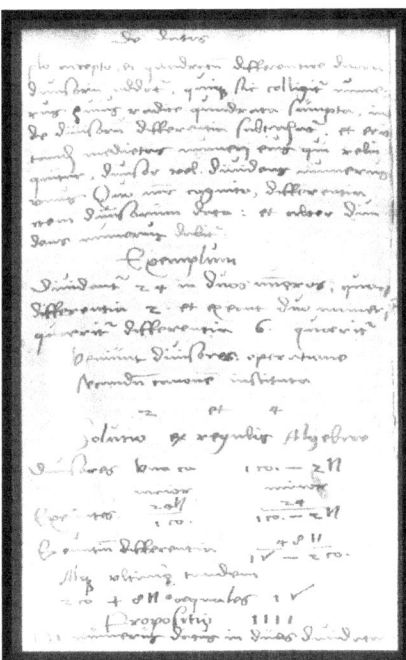

NL 173

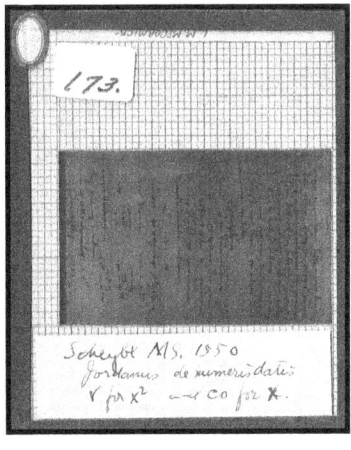

NL 174

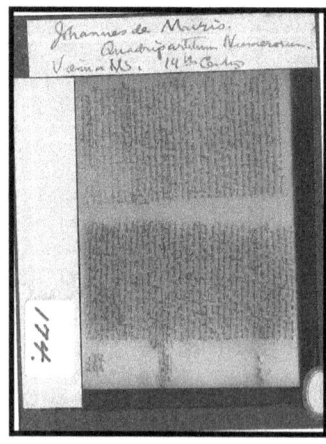

NL 175

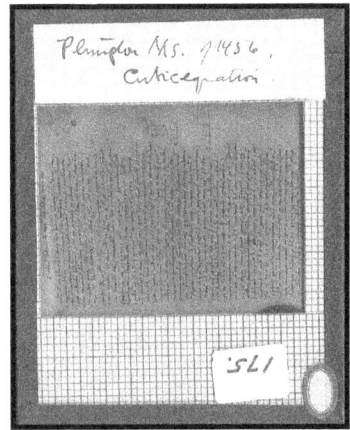

NL X

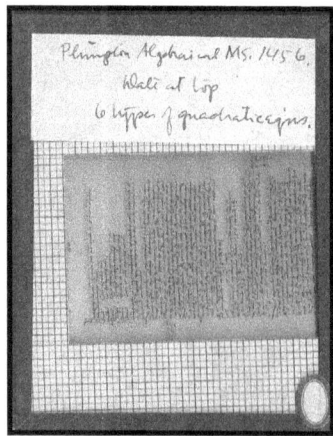

B. Lantern Slides

Checking teacher
"I just know she's wrong."

NL Y

298 B. Lantern Slides

Appendix C

The Educational Museum of Teachers College and the Department of Mathematics Pamphlets

Transcribed and reprinted under permission from the Rare Book and Manuscript Library, Columbia University.

Pamphlet #1

The Department of Mathematics extends to all teachers visiting the University an invitation to inspect and make use of the material available for the study of the Teaching and History of Mathematics in Teachers College. In addition to Professor Smith's library of several thousand books and pamphlets upon this subject, there is also available his collection of mathematical instruments – some dating as far back as 1450 – of manuscripts, and of engravings and portrait medals of eminent mathematicians.

This material may be examined in the Mathematical Library,

Room 212, Teachers College, the room being usually open, except on holidays, from 9 to 12 daily and from 2 to 4 daily except Saturdays.

There is also exhibited in Room 211, adjoining, a collection of mathematical apparatus and models adapted to the needs of the various grades from the Kindergarten through the High School. This includes number games, mensuration blocks, and models usable in geometry and trigonometry.

Early Mathematical Instruments. The early mathematical instruments exhibited included the following:

An Astrolabe of Arabic workmanship.

An Astrolabe of Italian workmanship, signed by the maker, and dated 1509.

An Astrolabe, a part dating from about 1450, and the rest, including the four plates, from the following century.

An Astrolabe of Paduan workmanship, signed by the maker, and dated 1557. A practically perfect specimen, with five finely engraved plates.

A Quadrans of the 16th century, one of the primitive instruments of trigonometry, bearing the early names "Umbra recta," and "Umbra versa."

Several Levelling Instruments of the 17th and 18th centuries.

Numerous measures of length and weight, of the 17th and 18th centuries, including the ell and some interesting sets of money changers' weights.

Several finely engraved Protractors, Diagonal Scales, and similar instruments.

Several Sector Compasses and compasses of other kinds, of the Renaissance period.

A collection of typical forms of Dials to illustrate the application of mathematics to dialing in the Renaissance period.

Several armillary spheres of the 16th, 17th, and 18th centuries.

C. Pamphlets

Medal Portraits of Mathematicians. This collection includes more than a hundred medals and medallions. The following are among the most prominent mathematicians represented:

Arago, Fr.	Huygens	D'Alembert Pascal
Archimedes	Kepler	De Moivre Pestalozzi
Aristotle	Lacroix	Descartes Poinsot
Bailly	Lagrange	Euler Poisson
Betrand	Lalande	Fermat Pythagoras
Bonnet	Laplace	Galileo Quetelet
Brahe, Tycho	LeVerrier	Gassendi Stevin
Cardan	Lobachevsky	Gauss Thales
Cassini	Maurolicus	Grandi Viviani
Cauchy	Monge	Halley Wolf
Cavalieri	Neudorffer	Hutton Wren
Copernicus	Newton (7)	

The complete set of mathematical portrait medallions by David d'Angers is included.

In addition to the portraits there are numerous other medals of interest in the history of mathematics, including the rare Metric System piece of 1872.

Portraits of Mathematicians. There are in Professor Smith's library about two thousand portraits of mathematicians. Of these it is possible to exhibit only a relatively small number. About forty are framed and can readily be examined, and if visitors wish to examine others in the collection they will be assisted in doing so.

The collection represents the work of a number of years and the repeated examination of the stocks of many European dealers. It is particularly rich in the works of early engravers, although containing a considerable number of photographs and modern process portraits. Reproductions of a number of the portraits have been made for school and college use, by The Open Court Publishing Co. of Chicago.

The collection of Newtons includes all of the most important portraits of this great mathematician and physicist. An effort has also been made to acquire all of the best portraits of Leibnitz, Descartes, Euler, the Bernoullis, Legendre, Monge, Cauchy, and others who stand out as particularly prominent in the creation of pure mathematics. The collection also includes the portraits of many who have achieved success in the field of applied mathematics, notably of men like Laplace, Lagrange, Huyghens, Bailly, and Arago.

Many of these portraits have been reproduced in stereopticon slides for the use of the department.

Autographs of Mathematicians. On account of the lack of space, it is possible to exhibit only a few of the more than two thousand autographs of mathematicians to be found in this library. The following are among the most interesting, and are shown in one of the wall cases:

Newton. A two-page manuscript demonstration written for one of his students at Cambridge.

Leibnitz. Autograph letter relating to a series of integrals.

Autograph letters of Sir William Rowan Hamilton, Euler, Johann Bernoulli, Mersenne (written about 1625), Maupertuis, Legendre, Wronski, and Arago.

Documents signed by Gauss, Laplace, and Lagrange.

Autograph letters from Poncelet to Liouville, Liouville to Direchlet, and Arago to Poncelet.

Autograph letters of the following mathematicians have been taken from the files so as to be accessible, and are usually displayed:

In pure mathematics: Jacobi, Cayley, Sylvester, Kronecker, Cremona, Hachette, Poincare, Hermite, Clebsch, Cauchy, Chasles,

C. Pamphlets 303

Clifford, Binet, Bezout, Monge;

In astronomy: Bode, Airy, Delambre, the three Cassinis, Maskeleyne, Flamsteed, Flammarion;

In physics: Ohm, Bessel;

In the history of mathematics: Montcula, Fuss, Libri, Kastner, P. Tannery, M. Cantor.

Miscellaneous Material bearing upon the History of Mathematics. There are also displayed a number of books and curios illustrating certain steps in the history or the teaching of mathematics. These include a Babylonian cylinder with cuneiform numerals, a piece of ancient Egyptian pottery with the zodiacal signs, Roman coins illustrating certain unusual forms in the ancient numeral system, some English tally sticks of 1296, two Renaissance computes medals, and a celestial sphere of the 16th century.

The bibliographical curios include one of the few copies saved from the fire which destroyed most of the first edition of Libri's "Histoire des Mathematiques" (vol. I), with Libri's autograph marginal notes. There are also autograph presentation copies of Laplace's "Theorie des Probabilites" and of Halliwell's "Rara Mathematica," over a hundred unpublished autograph letters of Prince Boncompagni on the history of mathematics, numerous first or early editions of works by such writers as Newton, Descartes, Tartaglia, Cardan, Bombelli, Paciuolo, Euler, and Barrow, a number of the earliest editions of Euclid, an unpublished French translation of Cantor's "Mathematische Beitrage zum Kulturleben de Volker," from the library of Chasles, and various similar works of bibliographical interest.

Mechanical Calculation. The material used to illustrate the development of mechanical calculation includes the following:

A collection of Mediaeval Counters (Jetons, Reckoning Pennies) of 15th and 16th century workmanship, partly French and partly German, some with the figure of the Rechenmeister seated at the abacus. Books showing the process of calculation by means of counters "on the line" will also be exhibited.

An Arabic abacus.

A Russian tschotu.

A Chinese swanpan.

A Japanese saroban.

A set of Napier's rods.

A set of Korean bones, the modern form of the ancient Chinese "Bamboo Rods," or the Japanese Sangi. Some Japanese books of 1698 will be exhibited showing the transition from this from of computing to the saroban, which took place in Japan about that time.

Modern calculating machines, including the Goldman and Stanley arithmometers, slide rules, and similar devices.

There are also available for study, in addition to those displayed, several early treatises showing the use of counters, together with numerous works on the historical development of this phase of arithmetic. This is also extensively illustrated in a collection of stereopticon slides belonging to the department.

Newtoniana. There are five framed portraits of Newton, as follows: Mezzotinit by Simon, after Thornhill; line engraving by George Vertue, after Vanderbank; line engraving by Houbraken, after Sir G. Kneller; lithograph by G.B. Black, after Wm. Gandy; line engraving by E. Scriven, after Vanderbank.

There are seven medals of Newton, representing the work of Croker (bronze and silver), Dassier, Roettiers, and Petit (two specimens), besides one without the artist's name.

The Newton manuscript was long in the library of Professor Jacoli, at Venice. It consists of a physical demonstration written by Newton at Cambridge, for an Italian student, c. 1700.

The impression of Newton's Galileo seal is from the original which was recently presented to the South Kensington Museum.

The bust of Newton is after the orginal by Roubillac.

The unframed portraits, numbering over one hundred, include specimens of the work of the following engravers: Phillibrown, Zeelander, Lips, Romney, Fry, Rivers, Scott, Tardieu, Ridley, Goldar, Cars, Laderer, Le Coeur, Freeman, Seeman, Krauss, Ravenet, Guadagnini, Holl, McGahey, Conquy, Zuliani, Cooke, Le Keux, Normand, Landon, Baumann, Wedgwood, Dupin, Smith, Edwards, Desrochers, Weber, and others.

Pamphlet #2

Educational Museum
Exhibition of Material Illustrating the Historical Development of Mathematics
From the collection of
David Eugene Smith, Ph.D., LL.D.
Professor of Mathematics
In Teachers College

The Educational Museum takes pleasure in announcing an exhibition of material illustrating the historical development of mathematics, from the collection of Professor Smith, beginning on Monday, January 4, 1909, and closing on Saturday, February 13, 1909. The Museum will be open on week days from 9 A.M. to 4:30 P.M., except on Saturdays when it will be closed at noon.

For the special benefit of teachers of mathematics in New York City and vicinity, Professor Smith will be present to explain the exhibit on Saturday morning, January 9, from 10:30 to 12.

The exhibit consists of mathematical instruments, measures, medals, manuscripts, early printed books, portraits, and curios, collected in various parts of the world and illustrating the history and teaching of mathematics in various periods. There are also photographs of many rare manuscripts and early printed works in various libraries of Europe and America, supplementing the original material in the collection.

For the benefit of visitors the following brief description of the contents of the various cases has been prepared, but all objects are carefully labeled.

Case I: Trigonometry and Astronomy

Books and instruments illustrating the early work in surveying, measuring of distances, and astronomy. The Renaissance quadrant is a specimen of one of the best known of the medieval instruments. The brass celestial sphere is a good example of the 16th century Italian work. The old Japanese sphere is of Nagasaki workmanship of about 1600. The Japanese manuscripts on trigonometry and surveying are particularly interesting, artistically as well as mathematically. The Ramsden telescope of 1775 was an excellent instrument for its time.

Case II: Scales and Weights and Chinese and Japanese Mathematics

There are some well-known problems relating to weights, that have been found in mathematical books for centuries. The weights here shown have been selected with a view to illustrating these problems. There are some interesting nests of weights from various German towns, and several curious sets of goldsmiths weights from different parts of Europe. Some of the scales are also interesting from the standpoint of the study of the lever.

In this case are also some of the more important Japanese and Chinese mathematical classics. These include the great Chinese encyclopedia of mathematics published under the Jesuit influence in the 17th century; the first Chinese edition of Vlacq's table of logarithms; an early Chinese edition of Euclid; numerous Japanese manuscripts and printed works, and an early Manchu treatise on mathematical astronomy. There are about five or six hundred Chinese and Japanese works in the library.

Case III: The Measure of Time

The study of the calendar represented the chief mathematical interest of early Church Schools, among people of all religions. In this case are shown various forms of sun dials, calendar medals, 16th and 17th century calendar rolls from the Buddhist temples of Japan, kalendaria from Europe, and three of the earliest publications on the Gregorian reform of the calendar.

Case IV: Medals of Mathematicians

An extensive collection of medals struck in honor of mathematicians. Among them are twelve medals of Newton, five of Descartes, four of Fermat, twelve of Galileo, the rare medals of the elder and younger Neudorfers, and some specimens of the best modern French work as seen in the portraits of Bertrand, Arago, and Le Verrier. There is also the complete set of medallions of mathematicians by David D'Angers.

Case V: Compasses, Measures, Astrolabes

Sets of compasses from the Roman to the Renaissance times. Sector compasses of various forms, protractors, and diagonal scales. Early measures of length and gaugers' scales.

The astrolabes and the armillary spheres include Italian pieces of the 15h and 16th centuries, Hindu, Persian, and Arab specimens, and some with an interesting history. There is one, for example, used by Pandit Joshti in restoring the observatory at Jaipur. It rests on a manuscript copy of the treatise on the astrolabe by the Maharajah Jey Sing.

Case VI: The Development of Number Systems

In this case is shown some of the material available for the study of the growth of various number systems, including the Hindu, Arabic, Roman, Greek, and Chinese. The Coptic manuscripts and Roman tesseras are particularly important.

Case VII: Number Mysticism and Games

This case contains a beautifully written manuscript, on silk, of the Yih King, one of the greatest Chinese classics, in which is found the first trace of the magic square, of permutations, and possibly of binary numerals. Both the magic square and the mystic trigrams are shown in various works in the case, and on numerous medals and amulets. The development into astrology is also shown, with some interesting Pali and Singhalese manuscripts on palm leaves.

In the right side of the case is shown the historical development of one of the oldest number games, that of dice, now at least three thousand years old, and the instructor in elementary number of more people than have ever learned counting in the schools. Dice from Etruscan tombs, from the remains of the Persian invaders, pre-Christian glass pieces from Karnak, a divinations icosahedral piece of the Ptolemaic period, the long pieces of the Roman conquerors of lower Egypt, loaded pieces from Rome itself, and so on down to the Renaissance period. There are between sixty and seventy pieces in the collection, dating from about 500 B.C. to the 18th century A.D., representing all varieties of marking from the typical Etruscan to the modern.

Case VIII: Mechanical Calculation

The development of mechanical calculation from the slates and possible abaci of the Neolithic age in Egypt to the modern arithmometer that divides one number by another by merely turning a crank. A squeeze of the Salamis abacus (the oldest one known), the Chinese swan pan, the Japanese soroban, Korean bones, the old Japanese sangi, the Russian stchotu, the Armenian abacus, and other similar forms are shown. There are also tally sticks dating from 1296, medieval counters placed on a line abacus, Napier's rods, and various later types of mechanical calculation.

Case IX: Rare Books

A few of the rare books from the library are here shown. They include several early European works on mathematics, while a few others are placed in Cases II and XIV. One of the half dozen copies of Libri's Histoire des Mathematiques, Vol. I, saved from the fire that consumed the rest of the first edition, is also shown.

Case X: Persian, Arabian, and Sanskrit Classics

A few of the manuscripts of the mathematical classics of Persia, Arabia, and India are here shown. There are in the case between two and three hundred manuscripts in these languages. Among the most interesting pieces in the library are several manuscripts of works by the greatest of the Hindu mathematicians, Bhaskara, who lived in the twelfth century, and these are shown in Case XXV.

Cases XI—XII: Illustrations for the Rara Arithmetica

The original photographs from which illustrations were made for Professor Smith's Rara Arithmetica. The works photographed are all in the library of George A. Plimpton, Esq., of New York. They included between three and four hundred arithmetics published before 1601, the largest collection that has ever been brought together.

Case XIII: Mathematics of Babylon

Casts of the mathematical tablets found by Professor Hilprecht at Nippur, and including those in the Imperial Ottoman Museum at Constantinople, with numerous illustrations from Professor Hilprecht's works. Two original cylinders with cuneiform inscriptions are also shown.

Case XIV: The Influence of Euclid

Arabic manuscript of Euclid written about 1300 A.D., together with a later copy, c. 1650. Manuscript of Matteo Ricci's Chinese translation of Euclid, written c. 1600. Several editions of Euclid of the 15th and 16th centuries. Several miscellaneous manuscripts on mathematics are also shown in this case, others being exhibited in Case XXVI.

Case XV: Mathematics of Egypt

Fac-similes of the Ahmes and the Akhmim papyri, the former being the oldest extant manuscript on mathematics, dating from c. 1700 B.C., and copied from one of c. 2300 B.C.

Case XVI: Portraits and Illustrations

The reproduction of Dürer's Melancholia to the left, shows the oldest magic square known to exist in print.

Case XVII: Portraits of Mathematicians

A few portraits of eminent mathematicians from a collection numbering over two thousand. This collection contains, for example, about one hundred and fifty portraits of Newton.

Cases XVIII—XXIII: Autographs of Mathematicians

A few autographs from a collection numbering over two thousand. There are shown letters and manuscripts of Newton, the Bernoullis, Laplace, Lagrange, Legenre, Gauss, Halley, Flamsteed, Mersenne, Euler, Bessel, Dupin, Cauchy, and many others who have helped to make the science what it is to-day.

Case XXIV: Rare Pamphlets

Dissertations of famous mathematicians, and rare memoirs and presentation copies.

Case XXV: The Bhaskara Manuscripts

Manuscripts of the mathematical works of Bhaskara, the greatest of native Hindu mathematicians. He wrote at Ujjain, and his Lilavati and Bija Ganita are known all over India, Ceylon, Persia, and other adjacent countries. These manuscripts range in date from c. 1400 to modern times. Other manuscripts of Bhaskara's works are shown in Case X.

Case XXVI: Manuscripts

A few miscellaneous manuscripts on mathematics, including an unpublished life of Galileo. Other similar manuscripts are shown in Case XIV.

There are shown about the walls a number of casts of early inscriptions from Chittagong, India, containing magic squares. The magic square is also illustrated in Case VII.

Owing to the lack of room in the museum it is impossible to exhibit a great many books and objects that should supplement what is here displayed. These include books showing the early history of the calculus and analytics, portraits, autographs, photographs of rare inscriptions, and illustrations of primitive instruments in addition to those in Cases I, III, and V.

Appendix D

List of Portraits in Smith's Collection

The following list of names is from the finding aid available at the Rare Book and Manuscript Library, Columbia University. The organization of the list is alphabetical, as it is in the Library. The formats of the portraits in this collection include: negatives, engravings, busts, medals, paintings, sketches, photographs, and bas-reliefs.

Abbe, Ernst
Abel, Niels H.
Abelard
Accum, F. Christian
Adams, John Couch
Adrain, Robert
Aenae, H.
Agassiz, Louis
Agnesi, Gaetana
Agostino, Baptista di
Agrippa, Henri Corneille
Agrippa, Marcus

Ahrens, W. E. M. G.
Airy, George Biddel
Alciatus, Andreas
Alcott, Amos Bronson
Alfonse the Tenth
Allamand, Jean N. S.
Allen, Thomas
Allman, George S.
Alsop, John
Amaldi, Ugo
Amodeo, F.
Ampere, Andre Marie

Anacharsis,
Anaxagoras
Anderson, D. M.
Anderson, John
Angelis, Steffanus
Anich, Petrus
Apianus, Petrus
Apollonius
Arago, Francois J. D.
Arago, Jacques
Araldi, M.
Aratus
Archimedes
Archytas
Arenhet, E.
Argol, Andreas
Aristophanes
Aristotle
Arkwright, Sir Richard
Arnauld, Antoine
Aston, Francis William
Astorini, F. Elia
Astronomy (Allegorical)
Aubert, Alexander
Auwers, Arthur von
Babbage, Charles
Bacon, Francis
Bacon, Roger
Bagwell, Gulielmi
Baier, Johann W.
Bailey, Francis
Bailly, Jean Sylvain

Ball, Robert
Ball, W. W. Rouse
Ballantine
Baltzer, Richard
Banfi, Johannes
Banus, J.
Barlaam, Calabrese
Barozzi, Jacopo
Barreme, Francois
Barrow, Isaac
Bartholomae, Sounvers
Bartoli, Cosimo
Basedow, Johann B.
Bassus
Battagline, Guiseppe
Bauernfeind, Carl M.
Bayle, Pierre
Beckmann, Johann
Bede, Venerable
Beethoven, Ludwig van
Belgrado, Jacopo
Bell, Eric Temple
Bellows, C. F. R.
Beltrami, Eugene
Beman, Wooster W.
Benel, F. W.
Beneventano, Marco
Berkeley, George
Bernardinus
Bernardus, Jacobus
Bernoulli, Daniel
Bernoulli, Jacques

Bernoulli, Jean
Bert, Paul
Bertius, Petrus
Bertrand, Joseph
Bertrand, Louis
Besant, W. A.
Bessel, Friedrich Wilhelm
Bessemer, Sir Henry
Betti, Enrico
Beyerus, Hartmann
Bezout, Stefano
Bianch, Luigi
Bickerdike
Bidder, George
Biddle, D.
Binet, M.
Bion, Nicholas
Biot, Jean B.
Birkhoff, George
Bjerknes, C. A.
Bjornbo, A. A.
Blackwood, Elizabeth
Blagrave, Joseph
Blagrave, John
Blake, James Gibbs
Blanchino, Francisco
Bliss, G. A.
Blissard, J.
Boaz, Franz
Bobart, Jacob
Bocher, Maxime
Bode, I.

Boehm, Andreas
Boerhaave, Hermann
Boetius, Anicius
Bohr, Niels
Boius, Johann
Bolyai, Farkas
Bonati, Teodore
Bonnet, Ossian
Booker, John
Booth, James
Borchardt, C. W.
Borelli, Giovanni
Borgesius, Johannes
Borromius, Alexander
Bortolotti, Ettore
Boscovich, Roger Joseph
Bossut, Charles
Boullaud, Ismael
Boulton, Matthew
Bowdick
Bowditch, Nathaniel
Boyle, Robert
Bradley, James
Brahe, Tycho
Brander, George F.
Brashear, John A.
Brechfeln, Christopher F.
Brewster, David
Brindly, James
Brioschi, Francesco
Brocard, H.
Brook-Smith, J.

Brooks, Edward
Brougham, Lord Henry
Brouncker, W.
Brown, Miss
Brown, Colin
Brown, Ernest W.
Bruno, Giordano
Bruyant, Nichlaus
Bryan, Mrs. Margaret
Buchan, David Stewart E.
Bude, Guillaume
Bunsen, Robert Wilhelm
Burali-Forti, C.
Burgatti, Pietro
Burkhardt, Heinrich
Burkhardt, Jean Charles
Burnside, W.S.
Busch, Johann
Bush, Vannavar
Butler, Nicholas Murray
Buxton, Jedidiah
Caesar, Julius
Cagnoli, Antonio
Cain, William
Cajori, Florian
Calkoen, Jan Fred
Calvin, Jean
Calvis, Seth
Campanella, Tommaso
Campensi, Albertus P.
Canovani, Stanislav
Cantor, Georg

Cantor, Moritz
Caramuel, Joahnnes
Caravelli, A. Vito
Carcavy, P. de
Cardan, Jerome
Carey, Mathew
Carneade
Carnot, L.M.N.
Carnot, Nicholas Leonard
Carr, G.S.
Carus, Paul
Casamia, Petrus G. P.
Casey, John
Cassini, Giandomenico
Catalan, Eugene Charles
Cauchy, Augustin
Cavalieri, Bonaventuro
Cayley, Arthur
Cellarius, Christopherus
Cellerier, Charles
Cellini, Benvenuto
Ceva, P.
Challis, James
Chamberlaine
Chamberlin, Thomas C.
Chandler, Charles F.
Chappe, Abbe d'Auteroche
Chase, Pliny Earle
Chasles, Michel
Chatelet, Emilie du
Chevalier, Charles
Christoffel, Elain Bruno

D. Portraits

Chrysippus
Chute, Horatius
Clairault, Alexis Claude
Clark, Edwin
Clark, Samuel
Clarke, A. R.
Clausius, Rudolf Julius
Clanius, Christophorus
Clayton
Clebsch, Alfred
Clifford, William Kingdon
Clinton, De Witt
Coble, Arthur
Cocker, Edward
Cockle, James
Coffin, James
Colburn, Zerah
Cole, F. N.
Colenso, John William
Coley, Henry
Columbus, Christopher
Comte, Auguste
Condamine, C. M. de la
Condillac, Etienne E.
Condorcet, Marie
Confucius
Constable, Samuel
Cook, James
Cope, Thomas F.
Copernicus, Nicolaus
Coppenol, Lieven von
Cossali, Pietro

Costard, George
Cousin, V.
Cox, William
Craanen, Theodore
Cramer, Gabriel
Craponne, Adam de
Creilling, Johannes C.
Cremona, Luigi
Critonius, Jacob
Crockett, C. W.
Crofton
Crontone, Aristeo di
Crousaz, Jean Pierre
Culpeper, Nicholas
Curbasto, Gregorio Ricci
Curie, Marie
Curie, Marie and Pierre
Curtis, Edmund
Curtius, Sebastianus
Daimler, Gottlieb
D'Alembert, Jean le rond
Dalby, J. F.
Dalrymple, Alexander
Dalton, John
Dante
Darboux, Gaston
Darwin, George
Da Vinci, Leonardo
Davis, R. F.
Davis, William
Davy, Sir Humphrey
Day, Henry George

Day, Jeremiah
Dee, John
Delambre, Jean-Baptiste
Democritus
De Morgan, Augustus
De Parcieux, Antoine
Derham, William
Des Aguliers, Jean T.
Descartes, Rene
Dewey, John
Dickson, Leonard
Diesel, Rudolph
Digges, Thomas
Dillon, E. W.
Dilworth, Thomas
Diogenes
Dion of Syracuse
Dirichlet, G. Lejeune
Dodgson, Charles L.
Dollond, John
Dollond, Peter
Doppelmaier, Johann G.
Doria, Paolo M.
Dowling, L. W.
Drebbel, C.
Dresden, Arnold
Drury, Dru
Duhamel, Jean M.C.
Duhamel-Dumonceau, H.
Duillier, Nicolas
Du Maurier, George
Dupuy, L.

Durer, Albrecht
Dyson, Sir Frank [Watson]
Eddington, Arthur
Einstein, Albert
Eisenhart, Z. P.
Eisenlohr, Friedrich
Eisenstein, Ferdinand G.
Eliot, Charles W.
Elliott, Edwin B.
Ellis, Alexander
Elsworthy, Miss
Ely, Achsah M.
Elyot, Sir John
Empedocles
Enestrom, Gustave
Engel, Friedrich
Enriques, Federigo
Epicurus
Erasmus, Desiderius
Escott, Albert
Euclid
Euler, Leonhard
Everett, J. D.
Exley, Thomas
Eytelwein, F. A.
Faber, Jacob
Fabre, Jean Henri
Fabricius, Johann Baptista
Fagnano, Julius Carlos
Faille, Johannes Carlos
Faraday, Michael
Farny, A. Droz

D. Portraits

Faucett
Faulhaber, Johannes
Fehr, Henri
Feltre, Vittorino da
Fergola, Niccola
Fernel, Johann
Ferguson, James
Fermat, Petrus de
Ferracinal, Bartolommeo
Ferrol
Feurbach, Ludwig
Fibonacci, Leonardo
Fiedler, Wilhelm
Figarie, E.
Filalao
Fine, H.B.
Fine, Oronce
Fisher, Irving
Fiske, Thomas S.
Flammarion, Camille
Flamsteed, John
Flawel, Pernelle
Fludd, Robert
Folkes, Martin
Fond, A. J. Sigaud de la
Fontand, Padre Mariano
Fontenelle, Bernard
Forsyth, Andrew Russell
Fortey, H.
Foster, G. Carey
Foster, W. S.
Foucault, Jean B. L.

Fouquet, Nicolas
Fourcroy
Fourier, Joseph
Fowler, R. H.
Fox, Henry
Fracastorius, Hieronymus
Franceshinis, Francesco
Francoeur, Louis B.
Franklin, Benjamin
Frederick II, King of Prussia
French, John R.
Fresnel, Augustin
Frisi, Paolo
Frontinus, Sextus Julius
Fuchs, Emmanuel Lazarus
Fujisawa, R.
Fulton
Furtenbachy, Joseph
Fuss, Paul Henri
Galgenmair, George
Galileo
Galliers. F.
Galois, Evariste
Galton, Francis
Galvani, Luigi
Garnier, J. G.
Garvin, J. L.
Garicus, Lucas
Gassendi, Pierre
Gauss, Carl Friedrich
Gay-Lussac
Gaza, Theodore

Gehler, Johann Samuel Traugott
Geikie, Archibald
Gelder, Jacob de
Gemma, Reinerus Frisus
Genese, R. W.
George of Trebizoud
Gerbert, Sylvester II, Pope
Gerhard de Jode
Germain, Sophie
Gernerus, Johannes
Gervis, Henri
Gibbs, Walcott
Gilbert, Davies
Gile, David
Gilliss, James Melville
Ginsburg, Jekuthiel
Gioja, Melchiore
Giorgini, Gaetano
Gisze, George
Gladstone, William E.
Glaisher, James
Gocchen
Godward
Goethe, Johann
Goldenberg
Goldmayer, Andreas
Goode, George B.
Granville, Earl
Grashof, Franz
Grassmann, Hermann Gunthur
Gravesande, Guilielmus
Greaves, John
Greene, George S.
Greenstreet, W. J.
Greenwood, J. M.
Greer, W. R.
Gregory, David
Gregory, James
Gregory, Olinthus
Guccia, Giovanni B.
Guericke, Otto de
Guglielmini, Domenico
Guicciardini, Francesco
Guillaume, Charles E.
Guyot, A.
Hachette, Jean Nicolas Pierre
Hadamard, Rudolph W.
Haeckel, Ernst
Hagen, Rev. John G.
Hajek Z Hajku, Tadeas
Hall, Asaph
Hall, G. Stanley
Hall, Spencer
Halley, Edmund
Halstead, George Bruce
Hamilton, William R.
Hammond, J.
Hanlon, G. O.
Hanow, Michael
Hanumanta, Nan B.
Hariot, Thomas

D. Portraits 321

Harley, Rev. Robert
Harris, Johannes
Harris, W.
Harrison, John
Harrod, Benjamin M.
Hartgill (or Hartgyll), George
Hartle, Heinrich
Harvey, Gideon
Hassler, Ferdinand Rudolph
Hatton, Edward
Hauck, Guido
Haughton, Samuel
Hayaski, T.
Heard, Robert Lynn
Heath, Sir Thomas
Heath, Royal V.
Hedrick, E. R.
Heis, E.
Hellwig, Johann Christ Ludwig
Helmart, Hermann L.
Helmholtz, Hermann L.
Henderson, Archibald
Hendricks, Joel E.
Henry, Joseph
Hensel, Kurt
Heppel, George
Heraclitus
Herbart, Johann Friedrich
Hermite, Charles
Herschel, Caroline

Herschel, John, baronet
Herschel, William, Sir
Hesse, Otto
Hevelius, John
Hiernon, King of Syracuse
Hilbert, David
Hildericus, D. Edo
Hill, George William
Hind, John
Hipparchus
Hippocrates
Hirst, J. Arthur
Hirth, Friedrich
Hobbes, Thomas
Hobson, Ernest William
Hodgson, Jacob
Hofmann, August Wilhelm von
Hofmann, Ulric
Hogendijk, Steven
Holbein
Holz
Holzmuller, Gustao
Hooper, Franklin
Hopkins, Rev. G. H.
Hopkinson, J.
Hopps, W.
Houel, Guillaume
Houel, Jean Hubert
Howe, Elias
Hudde, Johannes
Hudson, C. T.

Hudson, W. H. H.
Humboldt, Alexander
Humboldt, William von
Huniades, Johannes
Hunt, Charles Warren
Hutton, Charles
Huxley, Thomas H.
Huygens, Christian
Hypatia
Iacopo, Vincenzio di
Iezeler, Christoph
Inaudi, Jacques
Indagine, Ionnes
Ingleby, C. Mansfield
Irving, W.
Isbister, Alexander Kennedy
Jackson, C. S.
Jackson, Sir Herbert
Jacobi, Carl Gustav F.
Jacobi, Ferdinand
Jacobi, J. G.
Jacotot, Joseph
Jacquier, Francois
Jenkins, Morgan
Jenner, Edward
Jollois, Jean-Baptiste
    Prosper
Jones, Sir William
Jordan, Camille
Joule, James P.
Junius, Ulricus
Jurdak, M. H.

Kaestner, Abraham Gotthelf
Kant, Emmanuel
Kapteyn
Karsten, W. I. G.
Kasner, Edward
Kealy, James A.
Keeny, Abner C.
Kelland, Philip
Kelvin, William
Keppler, Johann
Kersey, John
King, Clarence
Kircher, Athanasius
Kirchoff, Gustav
Kirkman, Thomas Penygton
Kirkwood, Daniel
Kitchin, I.
Klebs, Arnold C.
Klein, Felix
Knight, W. M.
Knilling, Rudolph
Knowles, R.
Koertenblok, Joanna
Konigsberger, Leo
Kowaleaski, Sophie
Kowalski, Marian
Krafft, George Wolfgang
Kratzer, Nicolas
Krazer, Adolf
Kroneker, Leopold
Krupp, Alfred
Kruse, Jurgen E.

Kummer, Ernst Eduard
La Caille, Nicolas Louis
Ladd, Christine
Lagrange, Joseph Louis
Lahire, Phillipe de
Lalande, Joseph Jerome
 Lefrancais de
Lambert, J. G.
Lame, Gabriel
La Mettrie, Julien de la
Lamor, Sir Joseph
Landau, Edmund
Landry, Etienne N.
Langley
Lansberg, Mathieu
Lansbergius, Philipius
Lao-tze
Laplace, Pierre
Laurent, P. J.
Laval, Gustaf de
Laverty
Lavinal
Lavoisier, Antoine
Leaderdorf, C.
Lebesgue, Henri
Leclerc, Sebastien
Le conte, Joseph
Leeuwenhoek, Anthony van
Lefevre
Legendre, Adrieu-Marie
Leibnitz, Gottfried Wilheim
Leigh, C. W.

Lemaitre, Abbe
Lemoine, Emile-Michael
 Hyacinthe
Lenoir
Lenzio
Leonardo of Pisa (Fibonacci,
 Leonardo)
Lescher, S.
Leslie, Sir John
Lessing, Gotthold
Leucipus
Leuneschlos, Johannes A.
Leuschner, Armin
Le Vaillantm Francois
Le Verrier, Jean Joseph
Levi-civita, T.
Lewis, E. P.
Leybourn, Gulielmi
L'Hopital, Guillame Francois
 Antoine
L'Huillier, Simon
Liagre
Lie, Sophus
Lilly, William
Lindelof, Lorentz Leonard
Lindsay, James L.
Lipschitz, R.O.
Littrow, F.F.
Lobatchefsky, Nicolas
 Ovanovitch
Locke, John
Lockyer, Sir Norman

Lodge, Oliver J.
Lombard, Pierre
Long, Roger
Lorentz, Hendrik Anton
Lorenzoni, Guiseppe
Loria, Gino
Loschmidt, J.
Louisa, Queen consort of
    Frederick William III,
    King of Prussia.
Love, A. E. H.
Loyson, Charles
    (Pere Hyacinthe)
Lubbock, Sir John
Ludemann, Johann
    Christophorus
Ludolf van Collen
Ludovici, Jacobus
Lullus, Raymundus
Luyts, Jan
Lycurgus
McAlister, Donald
McCasy, W. S.
McClintock, Emory
McColl, Hugh
McCormick, Cyrus Hall
McDowell, J.
MacFarlane, Alexander
Mach, Ernst
McKenzie, J. L.
Maclaurin, Colin

MacLaurin, Richard
    Cockburn
McLeod, Lyons
McMahon, James
Magini, Jean Antoine
Mairan, J.J. Dortous
Malabri, Domenico Antonio
Mallet, Allain Manesson
Malus, Etienne L.
Mandey, Venteri
Mann, Horace
Mannes, Henri
Mannheim
Marat, Jean Paul
Marchetti, Alessandro
Marcus, Aurelius Antonius
Mari, Abate
Marinonil, Giovanni
    Giocomo
Marius, Simon
Marsham, John, baronet
Martin, Artemus
Martin, Benjamin
Martin, Rev. Hugh
Martin, Jacques
Martini
Martino, Nicolo de
Mascheroni, Lorenzo
Maskelyne, Nevil
Mason, Max
Maudit, Antoine R.
Maupertius, Pierre Louis

D. Portraits

Maurolycus, Franciscus
Maury, Mathew Fontaine
Maxwell, James Clerk
Mayne, Johannes
Melanchton, Philipe
Mendoza y Rios, Jose de
Mercator, Gerhard
Merrifield, C. W.
Merrifield, John
Mersenne, Marin
Metius, Adriaan
Metzler, W. H.
Meusinier, G. A.
Meziriac, Claude-Gaspard
Michelsen, Johann
    Andreas Christian
Mikami, Yoshio
Miller
Miller, Hugh
Miller, Kelly
Miller, W. H.
Miller, W. J. C.
Millikan, Robert
Milne, W. J.
Milner, Issac
Minchin, S. M.
Minkowski, Hermann
Mirandola, Pico della
Mitobius, Burchardus
Mittag-Leffler, Magnus
    Gosta
Mobius, August Ferdinand

Moigno, Abbe de
Moissan, Ferdinand
Molk, Jules
Moll, Gerrit
Monck, H.S.
Monge, Gaspard
Monro, C. J.
Montanarius, Geminianus
Montforte, Antonio di
Montgolfier, Joseph
Montgolfier, Stephane
Montreal, Jean de
Montucla, Jean Etienne
Moon, Robert
Moore, Amie Henrietta
Moore, E. H.
Moore, Sir Jonas
Moore, R. L.
Moors, Henry Erskine
Morin, Jean Baptiste
Morley, Frank
Morley, Thomas
Moss, John Calvin
Moul, A.
Mouraud, Salih
Moxon, Joseph
Mukhopadhyay
Mulcaster, J. W.
Mulerius, Nicolas
Muller, Johann Henricus
Muller, Max
Munsterus, Sebastianus

Murphy, Hugh
Murray, Daniel Alexander
Muspratt, James Sheridan
Musschenbroek,
    Johan Joosten van
Musschenbroek,
    Pieter van
Musschenbroek,
    Samuel Joosten van
Napier, John, eighth
    laird of Merchistoun
Napier, John
Nernst, Dr. Walther
Neuberg, J.
Neudorffer, Johannes
Neumann, Franz
Newcomb, Simon
Newmann, Carl
Newton, Hubert Anson
Newton, Isaac, Sir
Niceron, Jean Francois
Nicholson, Peter
Nicolai, Giambattista
Nicomedis
Nieuwland, Pieter
Noether, Emmy
Norden, John
Nostradamus, Michel
Nott, Eliphlet
Nowell, Alexander
O'Cogne
O'Connell, P.

Oddis, Mutii
Ohm, Georg Simon
Olearius, Adam
Oliver, Mrs.
Oliver, James
Olmstead, Denison
Oltramare, Gabriel
Oppolzer, Theodor von
Oreagan, John
Oriani, Barnaba
Origanus, David
Orlandi, Guiseppe
Orsted, Hans Christian
Ortelius, Abraham
Osgood, William Fogg
Ostrogradski, Michel
Oswaldus, Erasmus
Otto, Nicholas August
Ougtred, William
Owen, J. A.
Owen, Richard
Pagan, Blaise Francois de,
    comte de Merveilles
Painleve, Paul
Palitzsch, Jean George
Palmer, C. I.
Pappus, Johannes
Parrish, Celestia
Parsons, Sir Charles
    Algernon
Pascal, Blaise
Pascal, Ernest

D. Portraits

Pasini, Claudio
Pasteur, Louis
Patin, Charles
Patot, Simon Tyssot de
Payne, Roger
Peirce, Benjamin Osgood
Peiresc
Perouse, Jean Francois Galaup
Pestalozzi, Johann Heinrich
Pettee, George D.
Petzvals, Joseph
Peuil, Cora
Pfaff, Johann Friedrich
Phillip, Irving
Phillips, Andrew W.
Piazzi, Guiseppe
Picard, Emile
Piccolomini, Alexandro
Pierce, Benjamin
Pighius, Albertus
Pignatari, Filippo
Pincherle, Salavatore
Pisanus, Leonardo (Fibonacci, Leonardo)
Pitiscus, Bartholomaus
Pittacus
Plancius, Petrus
Planck, Max
Plantin, Christoph
Plato Playfair, John
Plimpton, George A.

Plucker, Julius
Poincare, Henri
Poinsot, Louis
Poisson, Simeon-Denis
Polignac, Camille
Poncelet, Jean Victor
Poole, Gertrude
Porro, Francesco
Porta, Johann Baptista
Postel, Guillaume
Praalder, Laurens
Price, B.
Priestley, Joseph
Proctor, Richard A.
Ptolemy I, King of Egypt, called Soter
Ptolemy, Claudius
Pugliesse, Guiseppe
Pupin, Michael Idvorsky
Putnam, T.M.
Pythagoras
Quetlet, Adolphe
Ramanujan, S.
Ramus, Pierre
Rankine, W. J. M.
Rathborne, Aaron
Rawlyns, Richard
Rawson, R.
Rayleigh, Lord John William Strutt
Raymond, William G.
Read

Reaumur, Rene Antoine Ferchault de
Recorde, Robert
Redfield
Reeve, N. D.
Regiomontanus (Muller, Johann Henricus)
Reid, Thomas
Renaldinus, Car
Ressel, Joseph Ludwig Franz
Reynaud, Baron
Rennie, John
Renshaw, Alexander
Repsold, Hans Adolf
Reuterdahl, Arvid
Revillus, D. Didaco
Ribiere
Riccati, Giordano
Riccati, Jacopo
Riccati, Vincenzo
Ricchebach, Giacomo Canco
Richard of Wallingford
Richardson, R. G. P.
Rickman, John
Riemann, Georg Friedrich
Riese, Adam
Riquet, P. P.
Rittenhouse, David
Ritter, Ernst
Rivard, Francois
Roberts, Ralph
Roberts, Samuel
Roberts, William
Robson, H. C.
Rockwood, Charles Greene
Rogel
Rogers, W. H. H.
Rogers, William Barton
Rohanet, Jacques
Rollo, Celestino
Romaine, William
Romanes, George John
Romney, Charles, Earl of
Rontgen, Wilhelm Conrad
Root, Oren
Rossignol, Antoine
Rosius, Jacob
Rosse, Earle
Rossi, Gaetano
Rouch, Rev.
Roussaeu, Jean Jacques
Rowe, Joseph Eugene
Rowland, Henry Augustus
Ruggieri
Runge, Carl
Runkle, J. D.
Rush, Benjamin
Russell, Bertrand H. W.
Russell, Robert
Rutherford, Earnest, Lord
Rutherford, Dr. William
Rutter, Edward
Sabine
Sacro Busto, Jean de

D. Portraits

Sadler, G. T.
St.John, Charles E.
Salimo, degli Armati
Salmon, George
Sanderson
Sarpi, Paolo
Sarton, George
Saunderson, Nicholas
Savage, Thomas
Saverien, Alexandre
Sawyer, Rachel
Saxo, Petrus
Scaliger, Josephus
Scarborough, Charles
Scheibner, Wilhelm
Scherling Ernst
Schikerdus, Wilhelmus
Schiller, Johann Christoph Friedrich von
Schiaparelli, Giovanni
Schloilch, O.
Schmid, Erasmus
Schmidt, I. R.
Schonerus, Johannes
Schooten, Frans Jr.
Schooten, Frans Sr.
Schopenhauer, Arthur
Schoute, P. H.
Schuler, Werner Joseph
Schumacher, Heinrich Christian
Schwarz, Hermann
Schwenter, M. Daniel
Scott, A. W.
Scott, Charlotte A.
Scott, Michael
Segner, John Andrew von
Seitz, E. B.
Serret, Joseph Alfred
Sextus Empiricus
Shakespeare, William
Sharpe, J. H.
Sharpe, James W.
Sids, J.
Simmons, Rev. C.
Sircom, Sebastian
Skeath, Walter William
Slichter, C. S.
Sluse, Rene Francois
Smeaton, John
Smith, Aquilla
Smith, David Eugene
Smith, Edgar
Smith, John
Smyly, J. G.
Smyth, William
Snell, Rodolphus
Snell, Willebrordus
Snow, Ralph
Snyder, Virgil
Socrates
Soddy, Frederick
Solon
Somerville, Mrs. Mary

Somes
Sommerfeld, Arnold
Sorbon, Robert
Speer, William W.
Sperry, Elmer
Spinoza, Benedict
Spottiswoode, William
Stallo, John Bernard
Stark, Johannes
Stas, Jean
Stedler, Johann Sebastian
Steele, R.
Stefan, Joseph
Steggael, J. E. A.
Steiner, J.
Steinschneider, M.
Stephanus, Robert
Stephenson, George
Stephenson, R.
Stevin, Simon
Stodius, Johann
Stoeras
Stoflerus, Johannes
Stokes, George Gabriel
Stoops
Stott, Walker
Strabo
Stratico, Simone
Strickland, Agnes
Strong, Theodore
Straud, A.
Struensee, Karl August von

Strum, Charles
Struve, W.
Struyck, Nicolas
Sturm, John Christoph
Sturm, Leonhardus Christophoros
Sulzer, Johann George
Swain, Joseph
Swinburne, Charles Algernon
Swinden, J. H.
Sylvester, James Joseph
Symons, E. W.
Taisnier, Jean
Tait, Peter Gutherie
Takagi, Teiji
Talete
Tallyrand, Henri Count de Chalais
Tanner, H.W. Lloyd
Tannery, Paul
Tarleton, Francis
Tartaglia, Niccolo
Taylor, Brook
Taylor, Charles
Taylor, Henry Martyn
Taylor, Johannis
Tchebychef, Pafnutiy Lvovich
Tebay, L.
Teleford, Thomas
Telesius, Bernardinus

D. Portraits

Tempelhoff, George
    Friedrich von
Tennyson, Alfred
Terry, Thomas R.
Thales
Theophrastus
Thomas, D.
Thomasi, Thomas de
Thompson, Benjamin
    Count Rumford
Thomson, Elihu
Thomson, F. D.
Thomson, Sir Joseph John
Thomson, Sir William
Todhunter, Isaac
Tonstall, Cuthbert
Tonelli, Leonida
Torelli, Guiseppe
Torricelli, Evangelista
    di Gaspero
Townsend, Richard
Tredgold, Thomas
Trew, Abdias
Trincano, L. C. V.
Tucker, Robert
Turner, Sir William
Tyler, Harry Walter
Tyndall, John
Van Amringe, J. H.
Van de Vyver, A.
Vancouson, Jacques de
Vanhee, Pere Louis

Van Vleck, Edward B.
Varvignon, Piere
Vasilier, Alexander
    Vasilievitch
Vaulzard, J.L. de
Veblen, Oswald
Verbiest, Ferdinand
Vesalius, Andreas
Vespucius, America
Vico, Francisco de
Viete, Francois
Vieth, Gerhard Ulrich Anton
Vignoles, Alphonsus
Villette, Francois
Vitruvius, Marcus
Vivian, Vincent
Volta, Alessandro
Voltaire, Francois
Von Bach, Carl
Vorsselman
Wald, A.
Walenn, W. H.
Wallis, John
Walmsley, John
Wangerin, Albert
Ward, Beatrice
Ward, Isabella
Ward, Johannes
Ward, Seth
Waring, Edward
Warren, Samuel
Watson, Stephen

Watson, Sir William
Watt, James
Weber, H.
Weidel, H.
Weidelrus, J. F.
Weirstrass, Karl
Weinberg, Joseph
Weiss, Edmund
Wentworth, George A.
Wertsch, F.
West, Benjamin
Weyc, Hermann
Wharton, Sir George
Wheatstone, Sir Charles
Whewell, William
Whiston, Rev. William
White, Arthur
White, Henry S.
Whitehead, Alfred, North
Whitney, Mary W.
Whitworth, William Allen
Wichell, George
Wieleitner, H.
Wien, Wilhelm
Wilkins, John
Wilkinson
Williamson, B.
Willis
Willis, John
Willsford, Thomas
Wilson, Henry
Wilson, James M.

Wilson, John R.
Winckler, A.
Winthrop, John
Wohler, Friedrich
Wolf, Christian
Wolfers, –Ph.
Wolstenholme, Joseph
Wood, De Volson
Woodall, Herbert
Woodhead, A. L.
Woodhouse, W. S. B.
Woodward, R. S.
Worcester, The Marquis of
Worth, E.
Wren, Sir Christopher
Wright, Rev. R. H.
Wright, Thomas
Wroncke
Xenocrates
Xylander, William
Young, B. A.
Young, John
Young, Thomas
Yuan-Yuan
Zach, Franz von
Zanotti, Francesco Maria
Zannotti, Eustachio
Zenon
Zerr, G. B. M.
Zeus
Zimmermann, Robert
Zoroaster

Appendix E

Catalog of Smith's Collection of Mathematical Instruments

The following catalog was used during the 2002 exhibition, "'The Ground of Arts': Mathematical Instruments and Illustrated Books from the David Eugene Smith Collection" at the Rare Book and Manuscript Library of Columbia University. It is noted throughout the catalog which items were included in the exhibit. This catalog also includes the number of the item in a 1927 catalog of Smith's instrument collection.

E. Mathematical Instruments

2002 Cat. (1927 Cat.)	Description	Inventory Number	Box Number
1 (180)	Counter used in number work. Presented by Miss Rebecca J. Slaymaker of Lancaster County, Pa. This counter was found in a house built in 1807, the home of Miss Slaymaker's forefather, and was doubtless used by members of the household.	27-191	A3
2 (87)	Modern Chinese geomancer's compass. Canton, 1907.	27-194	D4
3 (86)	Modern Chinese geomancer's compass. Peking, 1907.	27-195	B3
4 (85)	Modern Chinese geomancer's compass. Peking, 1907.	27-196	B3
5 (1)	Armillary sphere. Italian workmanship of 1550, as indicated by the forms of the numerals. Wooden support of much later date.	27-197	Exhibit case
6 (8)	Sphere, bronze with stars in silver. Persian, dated 1055 Hegira, for 1645 A. D. The emperor Humayoun had as his chief astronomer on Haddad, and it was his grandson who, as the inscription states, made this globe.	27-198	Exhibit case
7 (12b)	Hindu astrolabe ca. 1900	27-199	C2

E. Mathematical Instruments

8 (7)	Celestial sphere, papier maché, about 300 years old, of Nagasaki period	27-200	Exhibit case
9 (33)	Sundial, universal equinoctial, of Austrian workmanship, c. 1700. Signed piece, bearing on the bottom directions for adjustment, and maker's name. "Elev. Pol/Augsburg Par/48 Cracau Pra/50 Leipsig 51/And Vogl."	27-201	B3
10 (32)	Universal equinoctial sundial, 1748. Made by Johann Willebrand of Augsburg.	27-202	B3
11 (31)	Universal equinoctial sundial of Tyrolean workmanship, c. 1650.	27-203	B3
12 (23)	Universal equinoctial sundial with level. Tyrolese, 17th century.	27-204	D4
13 (30)	Universal equinoctial sundial of Austrian workmanship, c. 1750.	27-205	B4
14 (22)	Sundial, universal horizontal and vertical south dial. Venetian workmanship, c. 1650 with noteworthy decoration in steel on the back	27-206	B4

15 (29)	Sundial, equinoctial, with compass. German workmanship, c. 1750. Signed, Johann/Schretteger in/Augsburg.	27-207	B3
16 (54)	Sundial, only part, made by Dunod of Düsseldorf, gold plated, 18th century. Signed, "Claude Dunod A Düsseldorf."	27-208	B4
17 (28)	Universal equinoctial sundial with compass. Signed, "L. Grasel." 17th century.	27-209	B3
18 (27)	Sundial, equinoctial, with compass and level arrangement. Gold plated on brass. 18th century. On the top "Krigner Varsaviae."	27-210	B4
19 (53)	Sundial of German workmanship, 18th century. On back, "Eleav Poli/Lisbon 39 Rom/42 Venedig 45 Wein/Augspu Munche/48 Nurnb/ Heidelb/Regensp. 49 Rig/ Moscau 57/ L.T.M."	27-211	B4
20 (26)	Universal equinoctial dial with compass. Made by Schretteger of Augsburg, c. 1750.	27-212	B3

E. Mathematical Instruments

21 (24)	Sundial, universal equinoctial with level arrangement. Bavarian workmanship, 18th century. On bottom, "Elev. Poli/Lisbon 39 Rom/42 Vened 45 Ofen/47 Augsp Munch/Salzby 48/ L T M/"	27-213	B3
22 (25)	Universal horizontal dial with compass. Can be set for various latitudes. Made by Langlois, Paris, 18th century.	27-214	B4
23 (45)	Sundial, horizontal and vertical south dial, of Austrian workmanship, c. 1700. Wood. On top of cover latitudes of cities printed on paper.	27-215	D4
24 (48)	Sundial, horizontal, with lines for marking the hours of sunrise and sunset. Signed and dated, "J.C. Strigelius/Creilsheim. fec./1742." Also on top, "Createrit hicce Dies, nescitur origo secundi/ An labor an requies: sic transit fabula mundi." With compass and level arrangement. Nürnberg.	27-216	B4
25 (47)	Horizontal sundial, German workmanship; dated 1824.	27-217	Exhibit case

E. Mathematical Instruments

26 (20)	Horizontal sundial. Arab-Hindu workmanship, with numerals used in and about Jaipur, India.	27-218	B4
27 (19)	Dial, horizontal solar and lunar (only part). Augsburg workmanship, 18th century. On top "Horizontal Solis &Lunaræ." On bottom, "Nicolaus Rugendas Augsp 48 Gr."	27-219	B4
28 (46)	Sundial, horizontal, on pivot, oriented by means of magnet. Made by Magwald, Berlin, 19th century.	27-220	D4
29 (18)	Portion of silver gilt sundial. Italian workmanship, with arms of noble family. 17-18th century.	27-221	B4
30 (17)	Cubical sundial of the 18th century. Bavarian workmanship. Horizontal and vertical. South, north, east, and west.	27-222	Exhibit case
31 (44)	Chinese vertical sundial, brass. Purchased at Peking, 1907.	27-223	D4
32 (52)	Elaborate sundial of German workmanship, early 19th century.	27-224	B5

E. Mathematical Instruments 339

33 (49)	Sundial, horizontal and vertical, south dial with hour lines. German workmanship. Ivory, 17th century. On under side of lid, "Wan mein Got will/ So ist mein zil dar/ auf Ich mich/verlassen will."	27-225	B4
34 (43)	Chinese vertical sundial. Purchased at Canton, 1907.	27-226	B3
35 (40)	Ivory sundial on the hemispherical principle. Purchased at Peking, 1907.	27-227	A8
36 (39)	Ivory sundial on the hemispherical principle. Purchased at Peking, 1907.	27-228	A8
37 (51)	Sundial. Modern Chinese	27-229	B5
38 (42)	Sundial, vertical, Chinese. Modern Canton piece.	27-230	B3
39 (38)	Sundial, Chinese-Japanese pocket dial on the hemispherical principle. Purchased at Nikko, 1907.	27-231	B3
40 (37)	Sundial, Chinese pocket dial on the hemispherical principle. Purchased at Peking, 1907.	27-232	B3
41 (36)	Sundial, Chinese pocket dial on the hemispherical principle. Purchased at Peking, 1907.	27-233	B3

42 (35)	Sundial, Chinese-Japanese pocket dial on the hemispherical principle. Purchased at Nikko, 1907.	27-234	B3
43 (34)	Sundial, Chinese-Japanese pocket dial on the hemispherical principle. Purchased at Kyoto, 1907.	27-235	Exhibit case
44 (41)	Sundial. Japanese pocket dial with lens, colored glass for observing the sun, compass and hemispherical dial. Purchased at Nikko, Japan, 1907	27-236	B3
45 (55)	Ivory sundial. 18th century. French	27-237	B4
46 (56)	Ivory sundial. 18th century. French	27-238	B4
47 (73)	Japanese surveying instrument. Early 19th century. It shows European influence.	27-239	D2
48 (74)	Japanese surveying instrument. Same as No. 73.	27-240	E1
49 (57)	Ivory sundial. 18th century. German	27-241	B4
50 (21)	Sundial of Bohemian workmanship, dated 1800. Maker's name Joan Engelbreht.	27-242	B5
51 (58)	Chinese sundial. 18th century.	27-243	E3

E. Mathematical Instruments 341

52 (9)	Celestial sphere. Hindu. 1640. Bronze with stars in silver.	27-244	Exhibit case
53 (60)	Sundial, Chinese. 19th century.	27-245	B4
54 (83)	Quadrant, Italian. Ivory with cover. Early 19th century.	27-246	A8
55 (88)	Compass, French. Used for surveying. 19th century.	27-247	D3
56 (59)	Chinese sundial. 18th century.	27-248	B4
57 (50)	Sundial. French. 18th century.	27-249	B3
58 (89)	Compass, instrument for finding the true north and south direction in plane-table work.	27-250	D3
59 (61)	Part of sundial. German workmanship.	27-251	B4
60 (11)	Undated Hindu astrolabe. Probably about 1750.	27-252	E4
61 (16)	Hindu astrolabe and quadrant combined. 18th century. The tube at the top took the place of the telescope.	27-253	E5
62 (15)	Hindu astrolabe and quadrant. 17th century. Similar to No.11. On the astrolabe side the important stars are inlaid in silver.	27-254	C2

63 (14)	Astrolabe. Italian workmanship of 1558. This remarkably preserved specimen of the complete astrolabe is signed twice on the edge, "Patavii Bernardinvs Sabevs Faciebat MDLVIII." "Patavii Apvd Bernardinum Sabevm."	27-255	Exhibit case
64 (13)	Astrolabe. Italian workmanship of about 1450 and 1525. An examination of the numerals show that some maker of about 1525 took an older instrument for the back.	27-256	C2
65 (10)	Ancient Arab astrolabe. Presented on behalf of the Rev. James L. Fowle, missionary in Cæsarea, Turkey.	27-257	D1
66 (N/A)	Astrolabe. Hindu, Jaipur, 18th century. "Mr. Plimpton's Astrolabe"	27-257a	Exhibit case
67 (12a)	Undated Hindu astrolabe. Probably about 1750	27-258	C2
68 (2)	Armillary sphere. Italian workmanship of the 17th century. Pivot hole and direction of degree marks indicate that it contained an alidade at one time.	27-259	C2
69 (3)	Armillary sphere. Italian workmanship of the 17th century	27-260	A7

E. Mathematical Instruments

70 (6)	Armillary sphere. Tyrolese workmanship, 17th or 18th century.	27-261	Exhibit case
71 (5)	Armillary sphere. Hindu. Purchased from the astrologer of the Maharajah of Jaipur. Jaipur, India, 1908.	27-262	C2
72 (4)	Armillary sphere. French workmanship. 18th century	27-263	A7
73 (76)	Diopter for use in plane-table work.	27-264	Missing
74 (75)	Surveyor's diopter. 19th century.	27-265	D2
75 (72)	Surveying instrument. German, early 19th century, made by Wiskemann, Meminger.	27-266	E10
76 (71)	Telescope said to have been made by Ramsden of London, the great maker of mathematical instruments about 1775.	27-267	D2
77 (147)	Part of a circle, brass, for measuring angles. German of about 1700. Evidently connected with a telescope or a surveying instrument.	27-268	C3
78 (80)	Level, possibly intended for cannon. A signed piece: "Dav. Beringer fecit." 18th (?) century.	27-269	A8

79 (81)	Level of the nature of No. 80 but more simple. Probably German workmanship of about 1650-1750.	27-270	A8
80 (77)	Level. Italian workmanship of the 18th century. Signed, "N.S."	27-271	A7
81 (82)	Quadrants. Tyrolean workmanship of about 1600. This was the common trigonometric instrument of the medieval times. It ordinarily hung on a nail driven in a four-foot rod, the hole being on the side opposite the "Numervs vmbrae versae." The hole at the end of this side is for the support of the plumb line. The hole in the center was used when the quadrant lay on the staff horizontally. The alidade (radius) is new. Pfusterthal, South Tyrol.	27-272	Missing
82 (84)	Square. Etched brass piece of German workmanship. 18th century. Purchased at Munich. Geometric square. German. 18th century	27-273	B3
83 (136)	Brass diagonal and trigonometric scale. German workmanship of 1733. It is signed by the maker, "Inventor Pappelt 1773."	27-274	B5

E. Mathematical Instruments 345

84 (170)	Japanese clock. 19th century.	27-275	E12
85 (171)	Japanese clock. 19th century.	27-276	E12
86 (196)	German mechanic's compasses. 18th century.	27-277	C2
87 (197)	German adjustable compasses with screw.	27-278	C2
88 (198)	German adjustable compasses with screw.	27-279	C2
89 (195)	German artisan's compasses with quadrant. Dated 1696.	27-280	Missing
90 (205)	German proportional compass of the 18th century.	27-281	C2
91 (199)	German iron compasses. 18th century.	27-282	A6
92 (200)	German mechanic's compasses. 18th century.	27-283	C1
93 (204)	Ancient Roman proportional compasses about the beginning of the Christian era.	27-284	C2
94 (201)	Chinese compasses, made of bamboo, 19th century.	27-285	A6
95 (202)	Ancient Roman compasses, about the beginning of the Christian era.	27-286	A5
96 (203)	Ancient Roman compasses, about the beginning of the Christian era.	27-287	C2
97 (169)	Japanese rice measure.	27-289	C1

E. Mathematical Instruments

98 (63)	Sundial curves with the signs of the Zodiac. Bohemian workmanship of 1531. Pencil drawing.	27-290	Missing
99 (62)	Sundial curves, Mainz, 1676. Engraved.	27-291	Missing
100 (193)	Japanese sangi sticks used in solving equations in the Old Japanese algebraic system. Purchased in Kyoto, Japan, 1907.	27-292	D6
101 (67)	Calendar on Dutch tobacco box, dated 1799.	27-293	D2
102 (191)	"Wee" arithmetical slips. Similar to Napier bones. Modern English manufacture.	27-294	A4
103 (178)	Small Japanese soroban. 1904.	27-295	A3
104 (175)	Child's abacus, modern.	27-296	D5
105 (181)	Goldman's adding machine with instruction booklet.	27-297	C1
106 (186)	Multiplier, calculator, Reichelt and Co. Round paper with scales for quick multiplying.	27-298	D6
107 (257)	Chinese fortune telling vase with the pa kwa symbols.	27-299	E6
108 (79)	French leveling instrument. Brass. 19th century.	27-300	C2
109 (187)	Slide rule. Areas and cubes. 19th century.	27-301	C1

E. Mathematical Instruments

110 (266)	Modern Buddhist number beads, similar to the rosary and related to the ancient abacus. Purchased at Mandalay, Burma, 1908.	27-302	A6
111 (208)	French draftsman's straightedge. Brass. Fine engravings on one side. 18th century.	27-303	A6
112 (265)	Modern Mohammedan finger beads, a relic of the rosary and abacus. Purchased at Constantinople, 1908.	27-303	C2
113 (65)	Old Japanese calendar board. One side marked "dai" (great) and is shown during the long months. The opposite side is marked "she" (small) and is shown during the short months.	27-304	D3
114 (267)	Ten-Jin, the prince, who according to tradition, introduced arithmetic into Japan. Shrines in his honor are common in Japan. Early 19th century. Wood.	27-305	E4
115 (190)	Napier bones, modern German. Ten bones in a cardboard box for multiplying and dividing.	27-307	Exhibit case
116 (66)	German calendar board. One revolving disc missing.	27-308	D3

117 (253)	Horn book, modern. Old form of reckoning board.	27-309	Missing
118 (192)	Korean computing rods (bones), the modern form of the ancient Chinese "Bamboo rods" which Japan discarded about 1700. Brought from Korea in 1896.	27-310	A3
119 (254)	Arabic (?) amulet, found at Karnakm, Egypt. It illustrates the degenerate forms of the magic square.	27-311	A4
120 (182)	Tally sticks, old English. These tally sticks date from about 1296. Found in the Chapel of the Pyx, Westminster. One of the small pieces in the lot there found bore the name of William de Costello, who was sheriff in 1296. The ancient English tallies were ordered burned in 1834 and it is said to have been owing to the extra fires made up for this purpose that the Houses of Parliament were destroyed.	27-312	C1
121 (189)	Napier bones. Modern French manufacture.	27-313	A2
122 (269)	Japanese astronomer. Grotesque ivory figure. Purchased at Nikko, Japan, 1907.	27-314	Missing

E. Mathematical Instruments

123 (252)	Horn book, Armenian. Presented by Professor Farnsworth.	27-315	Missing
124 (64)	Egyptian terra cotta piece, showing the signs of the zodiac as used by the Greek scholars in Alexandria.	27-316	A7
125 (255)	Amulet. Christian-Kabalistic. "Tetragramonton. Sadia. El Eloim Zebaoth..." and other Hebrew words separated from each other by signs of the cross. In the center a magic square 8 x 8.	27-318	A5
126 (179)	Japanese soroban. Purchased in Japan in 1904 and presented by Professor Richards.	27-321	D5
127 (69)	Perpetual calendar. Same as No. 68. Engraved arabesques. Diameter 47 mm.	27-322	A7
128 (68)	Perpetual calendar. German, 18th century. Two discs moving on a third. Fixed feasts, lengths of the months, position of the sun, length of the day, hours of sunrise, hour of sunset, length of the night, days of the week and month. Gold plated. Engraved landscape and arabesques. Diameter 38 mm.	27-323	Missing Jan 1959

129 (168)	German pedometer. 18th century.	27-324	C1
130 (258)	Hindu jewel case from northern India, with lock.	27-325	E9
131 (207)	A set of drawing instruments in a shagreen case. German workmanship of the 18th century.	27-326	C1
132 (176)	Chinese swanpan.	27-327	D6
133 (172)	Arab abacus. Brought from Armenia in 1903 and presented by Professor Farnsworth.	27-328	D5
134 (206)	Planimeter, early form, made by Ott, Kempten, Germany, 19th century. Signed "A. Ott. Kempten."	27-329	D1
135 (177)	Chinese swanpan. Purchased in China in 1904 and presented by Professor Richards.	27-330	D6
136 (173)	Russian abacus. Purchased in St. Petersburg in 1901.	27-331	C1
137 (271)	The Chinese philosopher, Lao-Tze. In his writings he refers to the "knotted cords" used in computation.	27-332	E6
138 (70)	Perpetual calendar. Same as No. 68. Diameter 51 mm.	27-333	A7
139 (174)	Modern abacus for teaching children.	27-334	E4

E. Mathematical Instruments 351

140 (188)	French slide rule. With half millimeter scale on back.	27-336	C1
141 (251)	Motar and Pestle bought in Nürnberg.	27-337	E8
142 (261)	Brush and ink holder bought in Canton in 1907.	27-338	A5
143 (108)	Roman weight. Age unknown.	27-339	A1
144 (264)	Linkage used in solving cubic equations. Modern.	27-340	C2
145 (263)	Linkage for straight line work. Modern.	27-341	A6
146 (160)	Brass gauger's scale, marked in hundredths of a foot. Nürnberg about 1700.	27-342	B6
147 (262)	Trammel for constructing the Conchoid of Nicomedes. Student's work of 1890.	27-343	A6
148 (144)	English protractor and diagonal scale. A signed piece made by Cox and Son, London.	27-344	B6
149 (135)	Diagonal scale, German, 18th century.	27-345	B5
150 (133)	Measuring rod, French, showing relation between the French and German measures. Signed by Langlois, Paris.	27-346	C3

E. Mathematical Instruments

151 (130)	Brass rule, probably Italian, 17th century. The rule gives distances exact to one hundredth of an inch.	27-347	B6
152 (134)	Diagonal scale, German workmanship giving the Paris and Rhenish feet. 18th or early 19th century.	27-348	B6
153 (156)	Sector compasses. 18th century, signed by Brière, Paris.	27-349	C3
154 (150)	Sector compasses. Signed piece, made by Butterfield, Paris.	27-350	B6
155 (151)	Sector compasses. 18th century, French.	27-585	C3
156 (152)	Sector compasses. 18th century, signed by Chapotot, Paris.	27-586	E2
157 (131)	Compasses and measuring rod. Italian workmanship of the 15th century. Interesting not only for workmanship but also for the comparison between the Roman and the modern units of linear measure.	27-587	E1
158 (149)	Sector compasses. Signed piece, made by Butterfield, Paris.	27-588	E2
159 (158)	Sector compasses. 17th century, Italian.	27-589	C3
160 (154)	Sector compasses. 18th century, French.	27-590	E2

E. Mathematical Instruments

161 (155)	Sector compasses. 18th century, French.	27-591	E2
162 (194)	German draftsman's compasses for drafting angles of 45, 40, and 27 1/2 degrees. 18th century.	27-592	A5
163 (157)	Sector compasses. 18th century, French.	27-593	C3
164 (153)	Sector compasses. 18th century, signed by Chapotot, Paris.	27-594	C3
165 (145)	Protractor, brass. French workmanship of the 18th century. Gives central angles for the various inscribed n-gons where n equals 3, ... 12.	27-595	B5
166 (140)	Brass protractor, German workmanship of about 1700, with baroque decoration.	27-596	B5
167 (141)	Protractor, German workmanship. Probably early 19th century.	27-597	B6
168 (142)	Protractor, German workmanship. 18th century.	27-598	B6
169 (137)	Part of an instrument for measuring heights. Consists of a protractor and ruler with divisions indicating the umbra versa. The hole in the center was used for the alidade.	27-599	E1

170 (146)	Protractor, signed piece made by Langlois, Paris.	27-600	B5
171 (143)	Protractor, beveled, probably 18th century.	27-601	B5
172 (138)	Protractor, brass, German workmanship of the 18th century. Gives central angles for the various inscribed n-gons, where n equals 1, 2, ... 12.	27-602	B6
173 (139)	Protractor, brass, German, probably 18th century.	27-603	B6
174 (122)	Nürnberg linked brass rule. About 1800.	27-604	B5
175 (119)	Bavarian foot rule, brass-linked rule. Early 19th century.	27-605	B5
176 (121)	German foot rule of the 18th century. Wood with brass inlays.	27-606	Exhibit case
177 (120)	Nürnberg foot rule. Early 19th century.	27-607	B6
178 (161)	Brass gauger's scale. Nürnberg. About 1765. Marked in inches and hundredths.	27-608	Exhibit case
179 (127)	Ruler and measuring rod. German workmanship, dated 1703. Interesting on account of "lines of metals" for lead and iron.	27-609	B6
180 (209)	French draftsman's instrument. Brass. 18th century.	27-610	A5

E. Mathematical Instruments

181 (164)	Gauger's scale of the year 1716. Signed "Louis 1716." Such scales were used in the Visierrechnung in Germany and gauging in Great Britain.	27-611	F1
182 (123)	Curious Nürnberg measuring rod. Signed F.I.S. and dated 1781. It gives Bavarian inch and foot and is interesting because of its symbols.	27-612	Too long, no box
183 (163)	Gauger's scale, of Welsh manufacture. Signed "G M G 1777."	27-613	Too long, no box
184 (166)	German cloth measure. Just preceding the metric system.	27-614	F1
185 (159)	German gauger's rod, divided into 12 parts, each subdivided into 12 parts. Wood with brass ends. Early 19th (?) century.	27-615	F1
186 (165)	Gauger's scale. English manufacture. The units of measure entered on it are: Hogshead, kilderkin, barrel, etc.	27-616	Exhibit case

E. Mathematical Instruments

187 (128)	Austrian measuring rod, the ell, of 1732. It bears the date "Anno 1732." The ell varied considerably in different cities. This one if a little less than 26 inches, more exactly 65.8 cm. This rod is interesting on account of the curious indications of fractional divisions.	27-617	C3
188 (162)	German gauger's rod. Wood. Early 19th century.	27-618	F1
189 (129)	The Bavarian yard. Considerably shorter than the English yard. Early 19th century.	27-619	F1
190 (132)	Nürnberg ell measure. Marked "Nürnberg 1718." Shows fractional divisions.	27-620	Exhibit case
191 (109)	Chinese money changer's scales, sealed several times. Purchased at Kyoto, Japan.	27-621	B1
192 (110)	Chinese money changer's scales, 19th century. Purchased at Kyoto, Japan, 1907.	27-622	A1
193 (111)	Scales, Chinese money changer's, 19th century. Purchased at Peking.	27-623	B1
194 (112)	Money changer's scale purchased at Peking, 1907.	27-624	B1

E. Mathematical Instruments 357

195 (94)	Nest of Tyrolean weights, seven weights.	27-625	A8
196 (95)	Nest of weights. Tyrolese. One of the official seals bears the date 1807.	27-626	Exhibit case
197 (90)	A nest of German brass weights. About 1700.	27-627	B3
198 (96)	Nest of Austrian weights of the 18th century, selected, like No. 97, to illustrate the ancient, "Problem of Weights." This is elaborately decorated and is one of the best specimens of the weightmaker's art of that period. It has as least ten official seals, one bearing the date of 1787.	27-628	Exhibit case
199 (98)	Nest of Tyrolese weights if about 1800. One of the official seals bears the date 1822.	27-629	A1
200 (97)	Nest of the Tyrolean weights of about 1700, bearing the official stamp of Jufstein. Selected to show the origin of the "Problem of Weights." It is made up of seven weights, said to be, respectively, 1, 2, 4, –, 64 quintal, the total being 1 lot or 1/8 pound.	27-630	A1
201 (92)	Nest of weights, brass, elaborately carved, Bavarian workmanship, 17th-18th century.	27-631	A8

202 (91)	Nest of German weights with seal. About 1700.	27-632	B3
203 (99)	Nest of weights, Tyrolean, c. 1800, bearing the official stamp of Kufstein. One seal bears the date 1805. Contains seven weights.	27-633	A1
204 (93)	Nest of weights. Florentine exchanger's set, 17th-18th century.	27-634	A8
205 (100)	Nest of weights. German. 18th century.	27-635	A1
206 (105)	Money changer's weights. Venetian of about 1750. Purchased to illustrate (1) the Renaissance problems in the "Chain Rule;" (2) the late use of the ancient Roman disc notation for fractions of the "as;" (3) the problem in exchange as given in the Renaissance arithmetics. The set is remarkably complete.	27-636	B1
207 (101)	Case of French weights, dated 1669.	27-637	B1
208 (102)	Money changer's weights, French, c.1750.	27-638	B1
209 (103)	Money changer's weights, German, c. 1800.	27-639	A2
210 (104)	Money changer's weights, probably German, 18th century.	27-640	B1

E. Mathematical Instruments 359

211 (118)	Chinese steelyard with wooden beam. Early 19th century.	27-641	B2
212 (115)	Steelyard. German coin scales. 18th century.	27-642	B1
213 (116)	Type of steelyard. Malabar. Movable fulcrum instead of movable weight.	27-643	B2
214 (117)	Steelyard. India. Early 19th century.	27-644	B2
215 (114)	Scales for weighing. Germany 18th century.	27-645	B2
216 (124)	Measuring rod, German, before the metric system.	27-646	C3
217 (125)	Measuring rod, English, 18th century.	27-647	C3
218 (126)	Measuring rod, British, 18th century.	27-648	B6
219 (113)	"Royal Improved Patent Balance." To weigh and gauge sovereigns an half sovereigns. About 1825.	27-649	B1
220 (106)	Chinese weights, 18th century.	27-650	A1
221 (107)	Wosiahedron, Greco-Egyptian of the Ptolemaic period.	27-651	A1
222 (268)	Ten-Jin, the prince, who according to tradition, introduced arithmetic into Japan. Bronze	27-652	E3

223 (270)	Shotoku Taishi, c. 600, Japanese prince, considered the father of Japanese arithmetic. He is shown with a soroban, which is an anachronism.	27-653	E8
224 (183)	Tally sticks indicating number of prayers by the Pilgrims at the shrine of St. Gugan, Barra, Ireland.	27-654	A2
225 (184)	Canadian tally sticks (two).	27-655	A2
226 (185)	Philippine tally sticks. One whole piece and two halves in paper box.	27-656	A2
227 (167)	French callipers, millimeter. 19th century.	27-657	A3

E. Mathematical Instruments

228 (148)	Sector compasses, English. First described by Galileo, 1606. Nearly a century ago Benjamin Pike, Jr., had a shop at 294 Broadway, New York. From this he issued, in 1848, a small book on mathematical instruments. In it he speaks of sector compasses as follows:*"The Sector*-Of all mathematical instruments that have been contrived to facilitate the art of drawing, there is none so extensive in its use as the sector. It is a universal scale. It not only contains the most useful lines, but also by its nature renders them of general application."	27-658	C3
229 (259)	Iron treasure chest. German, 17th century. Interesting for type of lock used.	27-659	N/A
230 (260)	Iron treasure chest. German, 17th century. Interesting for type of lock used.	27-660	E8
231 (78)	Pocket level. Early 19th century. Probably German origin.	27-661	C2
232 (171a)	Clockworks. 19th century. Gift of Dr. Mendelson, ca. 1900	N/A	C5

E. Mathematical Instruments

233 (171b)	Clockworks. Wooden works of an early American clock. Gift of Dr. Mendelson	N/A	D5
234 (256)	Fifteenth century lucky charm, Chinese. Inscription: 1. Cho me– longevity; 2. Fuki–wealth and nobility; 3. Kingoku–gold and jewels; 4. Mando–plenty house. The reverse side has zodiac-like symbols.	(256) 462	A4

E. Mathematical Instruments

235-306 (210-250)	A collection of dice of various historical periods, exhibited in this collection to show the development if one of the oldest number games known.	27-155 to 27-174, 27-176 to 27-190, 27-192 to 27-198, 27-214, 27-217 to 27-219, 27-221, 27-227, 27-230 to 27-232, 27-234, 27-238 to 27-241, 27-243, 27-244, 27-277 to 27-279, 27-319, 27-320, 27-637 to 27-644	A5 A4 E11
307 (N/A)	Scales. Small English scales. Mid 18th century. Gift of Dr. Mendelson	111b	A1
308 (N/A)	Modern Chinese Tally Stick. Used as "water money"	183a	A3

309 (N/A)	Coin of Brabant dated 1478, Arabic numerals	(281) 247 T.C.	A4
310 (N/A)	Roman Bone Styluses for writing on wax tablets. Note the flattened ends, for corrections and obliteration. First century A.D.	273	A4
311 (N/A)	Amulet. Magic square on reverse of medal showing Venus (contains errors)	254a	A4
312 (N/A)	Amulet. Chinese, with signs of the Zodiac. 3"diameter	256a	A3
313 (N/A)	Chinese Ink Slab. Alabaster. Modern	271b	A5
314 (N/A)	English Tally Stick. Dale Library of Weights & Measures	182a	C1
315 (N/A)	Magic Cube. Each of the six surfaces is a Magic Square, totaling 194 in every line, horizontal, vertical, diagonal. Also, every quarter totals 194.	N/A	C1
316 (N/A)	Cuneiform Tablet No. 322. "Pythagorean triangles". Clay tablet, incomplete, ca. 1900-1600 B.C. Old Babylonian cuneiform script.	N/A	C1
317 (N/A)	Cuneiform Tablet 322. "Pythagorean numbers", ca. 1900-1600 B.C.	N/A	C1

E. Mathematical Instruments

318 (N/A)	Chinese Scribe's Ink Slab. In tray for rubbing ink.	271a	C1
319 (N/A)	Brass Die in Box. Nürnberg. Paper weight.	282	C1
320 (N/A)	Chinese Inch Measure. Modern.	146a	B6
321 (N/A)	Hourglass. 20 min glass. Early 19th century. Gift of Dr. Mendelson	60a	B6
322 (N/A)	Astrolabe. Turkish bazaar work ca.1900	(12c) 199b	D1
323 (N/A)	Wooden Chinese sundial with compass. Indicates the 24 seasons used by the farmers, 19th century.	37a	B4
324 (N/A)	Steelyard. American Steelyard, 19th century. Gift of Dr. Mendelson	118a	B2
325 (N/A)	Scales. Chinese jeweler's scales. Contemporary. Made in Canton. Scale rod made of camel bone, according to maker's notice. Gift of Dr. Mendelson	111a	B2
326 (N/A)	Wooden ball (label is missing).	N/A	E6
327 (N/A)	Stove tile (label is missing).	N/A	E7
328 (N/A)	Metric weights (Fairbanks). Incomplete set.	288	E7

E. Mathematical Instruments

329 (N/A)	Octant. Made by Thomas Howard, Liverpool. Gift of Dr. Mendelson	283	E5
330 (N/A)	Compass. Italian. Dated 1780	N/A	E4
331 (N/A)	Object with missing label.	N/A	E2
332 (N/A)	Napier's rods. Ivory in wooden box. England, 18th century	N/A	Exhibit case
333 (N/A)	Cuneiform Tablet. Ca. 1900-1600 B.C. (Plimpton 322)	N/A	Exhibit case
334 (N/A)	Cylinder Seal. Ca. 2291-2255 B.C. (Cuneiform 46-4). Created in Mesopotamia over four thousand years ago during the Akkad period (2334-2154 B.C.)	N/A	Exhibit case
335 (N/A)	Astrolabe. Hindu, Jaipur, 18th century. "Mr. Plimpton's Astrolabe"	27-257a	Exhibit case
336 (N/A)	Sundial. Ivory. Horizontal & vertical. Top of cover has pin missing. Moon calendar reveals number of hours the moon lags behind the sun. Mae by Hans Tröschel in 1603. Nürnberg	N/A	Exhibit case

E. Mathematical Instruments

337 (N/A)	Egyptian. Thot, the Egyptian god who, according to Plato, introduced arithmetic into Egypt. From the tombs of Thebes. Miniature, carved in light-green stone.	(280) 469 T.C.	Missing
338 (N/A)	Number Game. A hexagonal prism, bone, with an ivory handle for twirling. Nürnberg, ca.1800	(250) 242	A5
339 (N/A)	Surveying instrument	N/A	E1
340 (N/A)	Object with missing label	52A	E2
341 (N/A)	Elipsograph	289a	C4
342 (N/A)	Elipsograph	289b	E12
343 (N/A)	Telescope	71a	C6
344 (N/A)	Sextant. 18th century. Signed "Dolland, London" – a famous family of instrument makers. Formerly used by Captain Gorham Bassett of Hyannis, Massachusetts.	285	C5
345 (N/A)	Elipsograph	N/A	G1

E. Mathematical Instruments

346 (N/A)	Abacus	172a	D5
347 (N/A)	Weights for early American clock	N/A	D5
348 (N/A)	Wall chart of Greek, Roman & Arabic number systems	N/A	D5
349 (N/A)	Leon Lalanne. Regle à calcul à enveloppe de verre.1854	N/A	D5
350 (N/A)	Chinese swanpan.	176a	C1
351 (N/A)	Measuring instruments	N/A	D6
352 (N/A)	Compass in a box	N/A	D6
353 (N/A)	Cast of Euler's medal	N/A	D6
354 (N/A)	Roman coins with labels 247-258 T.C.	N/A	D6
355 (N/A)	Chinese and Korean coins	N/A	D6
356 (N/A)	Counters	N/A	D6
357 (N/A)	Casts of Roman coins and seals, Henry VIII coin, Newton farthing 1793 T.C.	N/A	D6
358 (N/A)	Dialing Templates	287a	C7

E. Mathematical Instruments

359 (N/A)	Dialing Templates	287b	C7
360 (N/A)	Chinese slab of pitch (broken)	286	C7
361 (N/A)	Computation forms	287c	C7
362 (N/A)	Instrument	N/A	C7
363 (N/A)	Framed picture	N/A	C8
364 (N/A)	Slide rule	N/A	C8
365 (N/A)	Slide rule	187a	C8
366 (N/A)	Brick from the house of Sir Isaac Newton	N/A	C8

Appendix F

Countries Visited by David Eugene Smith

The following is a list of 73 places visited by David Eugene Smith. The one-page document is not dated but is signed by Smith. It is located in Box 72 of Smith's Professional Papers at the Rare Book and Manuscript Library.

Countries Visited

Provinces and islands are not included unless they are generally referred to as separate states or regions.

Canada	Colombia	Holland
United States	Ireland	Belgium
Guatemala	Scotland	Germany
Honduras	England	Denmark
Costa Rica	Portugal	Norway
Nicaragua	Spain	Sweden
British Honduras	Morocco	Finland
Panama	Algeria	Russia
Nova Scotia	France	Poland

F. Countries Visited

- Austria
- Hungary
- Chekoslovakia [sic]
- Yugoslavia
- Italy
- Bulgaria
- Turkey
- Greece
- Switzerland
- Danzig
- Syria
- Palestine
- Iraq
- Iran
- Transjordania
- India
- Burma
- Siam
- Sumatra
- Java
- Cambodia
- Indo-China
- Malaya
- Philippine Islands
- China
- Mongolia
- Manchukuo
- Korea
- Japan
- Ceylon
- Egypt
- South Africa
- Mozambique
- Libya
- Tunisia
- S. Kurdistan
- Aden (Hadhramut)
- Malay Peninsula
- Albania
- Montenegro
- Jamaica
- Bahama
- Cochin China
- Senegal
- Rhodesia
- Orange Free State

Index

Abraham, David, 18
American Mathematical
    Society, 114
    Bulletin, 13
Andrews, Benjamin R., 5, 47
Arago, 57
Archimedes, 28
Arts, Museums of Peaceful, 83
Aryabhatta, 21

Barnard, Frederick, 36
Batak Museum, 25
Beg, Ulugh, 19
Beman, Wooster Woodruff, 10
Bernoulli, Daniel, 63
Bhaskara, 60
Boethius, 56
Bombelli, 56
Boncompagni, Prince
    Baldassarre, 14
Bryson Library, 83, 91
Butler, Nicholas Murray, 35, 36,
    38, 104

Carroll, Lewis, 63
Casati, Paolo, 57

Children's Industrial
    Exhibition, 35
College of the City of New York,
    34
Columbia College, 35–38
Columbia University, 4, 10, 12,
    20, 89, 101, 107, 114
    Butler Library, 114
    David Eugene Smith, 4
    Friends of the Library, 114
    Rare Book and Manuscript
        Library , 2, 75

d'Angers, David, 62
Dale, Samuel S., 12
David Eugene Smith
    Professional Papers, 56
Digges, Thomas, 57
Do, Li Ma, 23
Dodge, Grace Hoadley, 33, 34
Dodgson, Charles Lutwidge, 63
Donoghue, Eileen, 4
Dutton, Samuel, 46

Euclid, 24, 60
Evans, Augusta J., 7

Fresius, Gemma, 57
Frick, Bertha M., 2, 12, 13, 105, 106, 115

Gale, Arthur, 91
Galilei, Galileo, 23, 57, 60
Gilman, Daniel C., 35
Ginsburg, Jekuthiel, 13, 114
Grinnell, 91

Hervey, Walter L., 37
Hobby Club, 55, 56, 83, 114
Horace Mann School, 38
Hunter College, 105
Huntington, James, 90

Industrial Education Association, 33, 35, 40

Jacoli, Ferdinando, 14
Jewett, Clara L., 11
John Hopkins University, 35

Karpinski, Louis Charles, 90
Kellogg, George S., 40, 44, 46, 47
Khayyam, Omar, 28, 60
Kitchen Garden Association, 33
Knilling, Rudolf, 10
Kunz, George F., 83, 114

l'Hôpital, Guillaume de, 61
Library Friends Associations, 101
Libri, Guillaume, 15, 56

Locke, Leslie Leland, 93
Low, Seth, 39
Luse, Eva May, 24

Madori, 24
Mann, Horace, 35
Mathematical Association of America, 91, 114
McAleer, Helen Jewett, 11
McClenon, R. B., 91
McFarlane, C. T., 54
McMurry, Frank, 46
McNeil, Weatha Gale, 4
Michigan State Normal College, 10
Mikami, Yoshio, 23, 93
Mitchell, E. A., 91
Monroe, Paul, 93
Museums of Peaceful Arts, 84, 89
Museums of the Peaceful Arts, 114

Napier, John, 57
Neilson, Sarah Mitchell, 54, 82, 91
New York College for the Training of Teachers, 37
New York Museum of Science and Industry, 89
Newton, Sir Isaac, 23
Niekerk, A. van, 106

Olschki, Leo S., 14

Pacioli, 56
Permanent Educational Exhibit, 81, 82, 89
Pisano, Leonardo, 23
Plimption, George A., 103–105
Plimpton, George A., 10, 23, 47, 55, 81

Rajanubhab, Prince Damrong, 26
Reeve, William D., 91
Refior, Sophia, 95
Ricci, Matteo, 24, 60
Rubaiyat, 28
Russell, James Earl, 34, 38, 40, 43, 45, 46, 79

Sabeus, Bernard, 65
Sanford, Vera, 106
Scientific American, 53
Shah, S. Bahadur, 18
Simons, Lao G., 13, 105
Simons, Lao Genevra, 13
Smith, David Eugene
    Professional Papers, 75
Snedden, David S., 47
South Kensington Museum, 23
State Normal School
    Brockport, 10
    Cortland, 9
Stoeffler, Johann, 57
Syracuse University, 9, 10

Taylor, Fannie, 11, 24

Teachers College, 4, 10, 18, 24, 30, 33, 34, 37–39, 41, 43, 54, 80, 95, 101, 107, 112
    Department of Mathematics, 48, 61, 82, 83, 89
    Educational Museum, 2, 3, 34, 40, 42, 44–47, 53, 75, 79–82, 84, 101, 114
    Mathematics Library, 48
Tonstall, 57
Tusi, El, 60

Union Theological Seminary, 35
University of Kansas, 91
Upton, Clifford B., 83, 91, 112

Vaāhamihira, 21
Vanderbilt, George W., 38
Vlacq, 57

Webb, General Alexander, 34
Williamson, C. C., 102, 104

Yalden, Ernest G., 83

Zankel Hall, 41

www.ingramcontent.com/pod-product-compliance
Lightning Source LLC
Chambersburg PA
CBHW071949220426
43662CB00009B/1065